TRANSLATIONS

OF

MATHEMATICAL

MONOGRAPHS

Volume 28

THE FUNCTIONAL METHOD AND ITS APPLICATIONS

by

E. V. VORONOVSKAJA

AMERICAN MATHEMATICAL SOCIETY

Providence, Rhode Island 02904

1970

МЕТОД ФУНКЦИОНАЛОВ И ЕГО ПРИЛОЖЕНИЯ

Е. В. ВОРОНОВСКАЯ

Министерство Связи СССР
Ленинградский Электротехнический Институт Связи
им. проф. М. А. БОНЧ-БРУЕВИЧА

Ленинград 1963

Translated from the Russian by
R. P. Boas

International Standard Book Number 0-8218-1578-4
Library of Congress Card Number 70-138816

Printed in the United States

PREFACE

This book can be used as a textbook for graduate students of mathematics and those preparing for examinations in the theory of functions, and in particular in the theory of best uniform approximation.

The material presented here may also be useful for engineers and students of radio and electrical engineering who are concerned with problems of the theory of filters, amplifiers, pulse technology, etc.

To make it easier for engineers to read the book, it contains a brief introductory section on the theory of moment sequences and the simplest theorems of functional analysis.

The first part of the book contains an exposition of the new theory which was announced in a series of papers by the author between 1934 and 1953; a detailed exposition has not been published previously. The second part illustrates the theory by solving a number of problems of Čebyšev type which could not be handled by classical methods.

The author hopes that the availability of this new method and its effectiveness in solving problems of various kinds will help graduate students in finding thesis topics, and engineers in the theoretical analysis of their experimental problems.

The author wishes to thank her student M. Ja. Zinger, and E. L. Rabkin, assistant in the department of mathematics of the Leningrad Electrotechnical Institute of Communication, for their help in preparing this book for publication.

Leningrad, 8 July 1963

TABLE OF CONTENTS

INTRODUCTION

§1. Moment sequences and Hausdorff's theorems

Let the function $\phi(x)$ be defined on the interval $[a,b]$. We divide the interval arbitrarily into n parts by taking $a = x_0 < x_1 < \cdots < x_n = b$, and let n become infinite while $x_{i+1} - x_i$ tends to zero.

The *variation* of $\phi(x)$ on $[a,b]$ [**1**] is the upper bound

$$\operatorname{Sup} \sum_{i=0}^{n-1} |\phi(x_{i+1}) - \phi(x_i)| = \operatorname*{Var}_{[a,b]} \phi(x).$$

If this bound is finite, $\phi(x)$ is called a function of *bounded variation* on $[a,b]$ [**1**].

It is clear that a function of bounded variation does not have to be continuous, but it evidently has only discontinuities of the first kind, and the set E of these points of $[a,b]$ is countable [**1**].

In the special case when $\phi(x)$ is nondecreasing on $[a,b]$ we have

$$\operatorname*{Var}_{[a,b]} \phi(x) = \phi(b) - \phi(a).$$

Consequently a bounded monotonic function is of bounded variation.

EXAMPLE 1. $\phi(x) = 0$ for $0 \leqq x \leqq 1$ and $x \neq c$, where $0 < c < 1$; $\phi(c) = 1$. Here $\operatorname{Var}_{[0,1]} \phi(x) = 2$.

EXAMPLE 2. $\phi(x) = 0$ for $0 \leqq x \leqq 1/3$; $\phi(x) = 1$ for $1/3 < x \leqq 2/3$; $\phi(x) = -1$ for $2/3 < x \leqq 1$ (such a function is called a step function):

$$\operatorname*{Var}_{[0,1]} \phi(x) = 3.$$

EXAMPLE 3. $\phi(x) = x \sin(1/x)$ on $(0, 2/\pi]$ and $\phi(0) = 0$. This function is continuous and bounded, but

$$\operatorname*{Var}_{[0,2/\pi]} \phi(x) = \infty.$$

EXAMPLE 4. $\phi(x)$ is a step function with jumps at the points $1/2, 2/3, 3/4, \cdots, n/(n+1), \cdots$.

1) If the corresponding jumps are $1/2^2, 1/3^2, \cdots, 1/(n+1)^2, \cdots$, then

$$\operatorname*{Var}_{[0,1]} \phi(x) = \sum_1^{\infty} \frac{1}{(n+1)^2} = \frac{\pi^2}{4} - 1.$$

2) If, however, the jumps are $1/2, 1/3, \cdots, 1/(n+1), \cdots$, then $\operatorname{Var}_{[0,1]}\phi(x) = \infty$.

Let $f(x)$ be continuous on $[a,b]$ and let $H(x)$ be of bounded variation. With arbitrary intermediate points $a = x_0 < x_1 < \cdots < x_{n-1} = b$, form the sum

$$\sum_{i=0}^{n-1} f(\xi_i)\,[H(x_{i+1}) - H(x_i)],$$

where $x_i < \xi_i < x_{i+1}$. If $x_{i+1} - x_i$ tends to zero as the number of division points is increased $(n \to \infty)$, the limit of the sums exists and does not depend on the choice of the points ξ_i in the intervals containing them.

This limit is called the *Stieltjes integral* [1] of $f(x)$ with respect to $H(x)$,

$$\int_a^b f(x)\,dH(x) = \lim_{n\to\infty} \sum_{i=0}^{n-1} f(\xi_i)\,[H(x_{i+1}) - H(x_i)]. \tag{1}$$

If $H(x) = x$ the Stieltjes integral reduces to the Riemann integral. Let $H(x)$ be a step function on $[a,b]$ with jumps of amount $\delta_1, \delta_2, \cdots$ at the points $\sigma_1, \sigma_2, \cdots$; then the Stieltjes integral reduces to a sum (or a series), since $dH(x) = 0$ except at the jumps. We obtain

$$\int_a^b f(x)\,dH(x) = \sum_{(i)} f(\sigma_i)\,\delta_i.$$

The definition of the Stieltjes integral implies that changing $H(x)$ at isolated points of $[a,b]$ does not change the value of the integral; and adding a constant to $H(x)$ does not change the value of the integral.

We also note the following obvious property of the integral: if $\max|f(x)| = A$ then

$$\left|\int_a^b f(x)\,dH(x)\right| \leq A \cdot \operatorname*{Var}_{[a,b]} H(x). \tag{2}$$

We can now define a moment sequence. If a sequence of real numbers is expressed in the following integral form,

$$\mu_n = \int_a^b t^n dH(t) \quad (n = 0, 1, 2, \cdots), \tag{3}$$

where $H(t)$ is a function of bounded variation, we call $(\mu_n)_0^\infty$ a *moment sequence* on $[a,b]$, and $H(t)$ its *integrator function* [1].

By a previous remark, a moment sequence is not changed if $H(t)$ is altered on a countable set of points of $[a,b]$, since these points can be avoided in forming the sums (1); also $(\mu_n)_0^\infty$ is not changed if $H(t)$ is replaced by $H(t) + C$. For definiteness we shall suppose that the integrator function $H(t)$ of a moment sequence satisfies

$$H(a) = 0 \quad \text{and} \quad H(c) = \frac{H(c+) + H(c-)}{2}$$

at each interior point c. Such a function is said to be normalized.

We leave open for the present the question of whether these conditions are sufficient for every moment sequence to determine its normalized integrator function uniquely (Corollary 2 of Hausdorff's Second Theorem) [2].

From now on we take the interval of integration for a moment sequence to be $[0,1]$, since this choice simplifies the formulas.

This restriction is not essential, since the linear transformation $t = a + x(b - a)$ carries the interval $[a,b]$ in t to the interval $[0,1]$ in x.

The problem of finding conditions under which a given sequence $(\mu_n)_0^\infty$ is a sequence of moments on $[0,1]$ involves investigating the differences of the numbers (μ_n), and leads to a different definition of moment sequences that is independent of their integral representation.

Put $\mu_n = \mu_{n,0}$ and $\mu_{m,n+1} = \mu_{m,n} - \mu_{m+1,n}$; then

$$\mu_{m,n} = \mu_{m,0} - \binom{n}{1} \mu_{m+1,0} + \binom{n}{2} \mu_{m+2,0} - + \cdots (-1)^n \mu_{m+n,0}.$$

In particular, if none of the differences $(\mu_{m,n})$ $(m = 0,1,2,\cdots; n = 0,1,2,\cdots)$ is negative, $(\mu_n)_0^\infty$ is called an *absolutely monotonic sequence.*

HAUSDORFF'S FIRST THEOREM [3]. *If a sequence of real numbers $(\mu_n)_0^\infty$ satisfies*

(4) $$M_p = \sum_{k=0}^{p} \binom{p}{k} |\mu_{k,p-k}| \leq L \text{ for } p = 0,1,2,\cdots,$$

i.e. if the sums M_p are bounded, then $(\mu_n)_0^\infty$ can be represented as the difference of two absolutely monotonic sequences $(\alpha_n)_0^\infty$ and $(\beta_n)_0^\infty$, i.e. $\mu_n = \alpha_n - \beta_n$ $(n = 0,1,2,\cdots)$.

We note first that

(5) $$\sum_{k=0}^{p} \binom{p}{k} \mu_{k,p-k} = \mu_0,$$

for any sequence, as is easily verified by induction. Hence if $(\mu_n)_0^\infty$ is absolutely monotonic, $M_p = \mu_0$ and (4) is satisfied.

Using the formula $\mu_{m,n} = \mu_{m+1,n} + \mu_{m,m+1}$ (for arbitrary m and n), noted above, we can write

$$\begin{aligned}|\mu_{m,n}| &\leqq |\mu_{m+1,n}| + |\mu_{m,n+1}| \\ &\leqq |\mu_{m+2,n}| + 2|\mu_{m+1,n+1}| + |\mu_{m,n+2}| \leqq \cdots\end{aligned}$$

a sequence of nondecreasing sums. The general form of one of these sums is

$$\sum_{h=0}^{p} \binom{p}{h} |\mu_{m+h,n+p-h}|;$$

we can estimate it as follows for all $m \geqq 0$ and $n \geqq 0$:

$$\sum_{h=0}^{p} \binom{p}{h} |\mu_{m+h,n+p-h}| \leqq \sum_{h=0}^{p} \binom{m+n+p}{m+h} |\mu_{m+h,n+p-h}| \leqq M_{m+n+p} \leqq L$$

(we need to use the fact that

$$\binom{p}{h} \leqq \binom{m+n+p}{m+h},$$

and put $m + h = k$).

Thus

$$\lim_{p\to\infty} \sum_{h=0}^{p} \binom{p}{h} |\mu_{m+h,n+p-h}| = \pi_{m,n} \tag{6}$$

exists. If we replace p by $p + 1$ and use the formula

$$\binom{p+1}{h} = \binom{p}{h} + \binom{p}{h-1},$$

we obtain

$$\begin{aligned}\lim_{p\to\infty} \sum_{h=0}^{p+1} \binom{p+1}{h} |\mu_{m+h,n+p+1-h}| &= \lim_{p} \sum_{h=0}^{p} \binom{p}{h} |\mu_{m+h,n+1+p-h}| \\ &\quad + \lim_{p} \sum_{h=1}^{p} \binom{p}{h-1} |\mu_{m+1+h,n+p-h}|.\end{aligned}$$

Thus $\pi_{m,n} = \pi_{m,n+1} + \pi_{m+1,n}$ and consequently the numbers $(\pi_{m,n})$ are the differences of

$$\pi_{m,0} = \lim_{p\to\infty} \sum_{h=0}^{p} \binom{p}{h} |\mu_{m+h,p-h}| = \pi_m,$$

and all $\pi_{m,n} > 0$. Therefore $(\pi_m)_0^\infty$ is absolutely monotonic; in addition, since

$$|\mu_{m,n}| \leqq \sum_0^p \binom{p}{h} |\mu_{m+h,n+p-h}|,$$

we have $|\mu_{m,n}| \leqq \pi_{m,n}$. Putting $\bar{\alpha}_m = (\pi_m + \mu_m)/2$ and $\bar{\beta}_m = (\pi_m - \mu_m)/2$, we have $(\bar{\alpha}_m)_0^\infty$ and $(\bar{\beta}_m)_0^\infty$ also absolutely monotonic, and $\mu_n = \bar{\alpha}_n - \bar{\beta}_n$. This completes the proof.

Corollary 1. *The sequences $(\bar{\alpha}_n)_0^\infty$ and $(\bar{\beta}_n)_0^\infty$ defined above are called the "minimal components" of $(\mu_n)_0^\infty$; this means that there is no absolutely monotonic sequence $(\gamma_n)_0^\infty$, except the zero sequence, that can be subtracted from $(\bar{\alpha}_n)_0^\infty$ and $(\bar{\beta}_n)_0^\infty$ without making one of them fail to be absolutely monotonic.*

In fact, the components that we have determined are such that the number L in (4) can be taken to be

$$\pi_0 = \lim_p \sum_{h=0}^p \binom{p}{h} |\mu_{h,p-h}| = \bar{\alpha}_0 + \bar{\beta}_0, \tag{7}$$

and this number evidently cannot be decreased.

For every other pair of absolutely monotonic components $(\alpha_n)_0^\infty$ and $(\beta_n)_0^\infty$ we have $|\mu_{m,n}| = |\alpha_{m,n} - \beta_{m,n}| \leqq \alpha_{m,n} + \beta_{m,n}$, and, by (5), $M_p \leqq \alpha_0 + \beta_0$. If we suppose that $\alpha_n = \bar{\alpha}_n - \gamma_n$ and $\beta_n = \bar{\beta}_n - \gamma_n$ then $\alpha_0 + \beta_0 = \bar{\alpha}_0 + \bar{\beta}_0 - 2\gamma_0$, which is possible only for $\gamma_0 = 0$. The uniqueness of our minimal components leads to the remark that not only are their termwise differences $\bar{\alpha}_k - \bar{\beta}_k = \mu_k$ determined, but so are their sums $\bar{\alpha}_k + \bar{\beta}_k = \pi_k$; in fact, $\bar{\alpha}_0 + \bar{\beta}_0 = \pi_0$, but the sequence $(\mu_n)_1^\infty$ has minimal components $(\bar{\alpha}_n)_1^\infty$ and $(\bar{\beta}_n)_1^\infty$ (the proof is word-for-word the same), and then the value $\bar{\alpha}_1 + \bar{\beta}_1 = \pi_1$ is minimal, etc.

Corollary 2. *We have shown that the sums M_p are monotone increasing (nondecreasing), and hence* (4) *is equivalent to the existence of*

$$\lim_p \sum_{k=0}^p \binom{p}{k} |\mu_{k,p-k}| = M. \tag{8}$$

Hausdorff's Second Theorem [2]. *A necessary and sufficient condition for a given sequence $(\mu_m)_0^\infty$ to be a moment sequence is that* (4), *or equivalently* (8), *is satisfied.*

1. Suppose that $\mu_m = \int_0^1 t^m dH(t)$. Form the differences

$$\mu_{m,1} = \int_0^1 t^m(1-t)\,dH(t),$$

etc.; in general, $\mu_{m,n} = \int_0^1 t^m(1-t)^n dH(t)$. Then, putting $\epsilon_{m,n} = \operatorname{Sgn} \mu_{m,n}$, we may write

$$|\mu_{m,n}| = \int_0^1 t^m(1-t)^n \epsilon_{m,n} dH(t)$$

and

$$M_p = \int_0^1 \sum_0^p \binom{p}{k} t^k(1-t)^{p-k} \epsilon_{k,p-k} dH(t);$$

but since

$$\left| \sum_{k=0}^p \binom{p}{k} t^k(1-t)^{p-k}\epsilon_{k,p-k} \right| \leqq 1$$

on $[0,1]$, it follows from (2) that $M_p \leqq \operatorname{Var}_{[0,1]} H(t)$. This establishes the necessity.

2. Suppose that $(\mu_n)_0^\infty$ satisfies (8). We shall obtain an integral representation for (μ_n). We introduce the abbreviation

$$M_p = \sum_{k=0}^p |\lambda_{k,p}|, \quad \text{where} \quad \lambda_{k,p} = \binom{p}{k} \mu_{k,p-k},$$

and construct a step function $H_p(t)$ on $[0,1]$: $H_p(0) = 0$; $H_p(t) = \lambda_{0,p}$ for $0 < t \leqq 1/p$; $H_p(t) = \lambda_{0,p} + \lambda_{1,p}$ for $1/p < t \leqq 2/p$; and so on. Then $H_p(t) = \sum_{k=0}^{p-1} \lambda_{k,p}$ for $(p-1)/p < t < 1$; $H_p(1) = \sum_{k=0}^{p} \lambda_{k,p}$.

We now apply Helly's First Theorem [**1**]: If the functions of the sequence $\{H_n(t)\}_1^\infty$ have collectively bounded variation on $[a,b]$, i.e. $\operatorname{Var}_{[a,b]} H_n(t) \leqq K$ for all n, it is always possible to select a subsequence $\{H_{n_k}(t)\}_{k=1}^\infty$ so that $H_{n_k}(t)$ converges at each point of $[a,b]$ to a function $H(t)$ of bounded variation on $[a,b]$.

Thus in our case, having constructed a sequence of step functions $H_p(t)$ whose variations on $[0,1]$ are bounded by M, we may select a subsequence so that $H_p(t)$ tends to some $H(t)$ (in general, not explicitly known) on $[0,1]$. Form the integrals $\int_0^1 t^m dH_{p_k}(t)$. By Helly's Second Theorem [**1**], under our hypotheses we can take the limit under the integral sign to obtain

$$\lim_k \int_0^1 t^m dH_{p_k}(t) = \int_0^1 t^m dH(t). \tag{9}$$

Since $H_p(t)$ is a step function, the integral on the left of (9) is a sum,

$$\int_0^1 t^m dH_p(t) = \sum \binom{p}{k} \mu_{k,p-k} \left(\frac{k}{p}\right)^m,$$

which, as we shall prove, tends to μ_m as $p \to \infty$ $(m = 0, 1, 2, \cdots)$ [4]. We have

$$\sum_{k=0}^{p} \binom{p}{k} \mu_{k,p-k} = \mu_0$$

for every p; consequently

$$\sum_{k=0}^{p} \binom{p}{k} \mu_{k,p-k}\frac{k}{p} = \sum_{k=1}^{p} \binom{p-1}{k-1} \mu_{k,p-k} = \mu_1$$

and generally

$$\sum_{k=0}^{p} \binom{p}{k} \mu_{k,p-k}\frac{k(k-1)\cdots(k-m+1)}{p(p-1)\cdots(p-m+1)} = \sum_{k=m}^{p} \binom{p-m}{k-m} \mu_{k,p-k} = \mu_m.$$

Now consider the expressions

$$\phi_m(k,p) = \frac{k^m}{p^m} - \frac{k(k-1)\cdots(k-m+1)}{p(p-1)\cdots(p-m+1)}$$

for $m \leqq k \leqq p$; clearly $\phi_m(k,p) \geqq 0$.

We have

$$\phi_m(k,p) = \frac{k^{m-1}(p-1)\cdots(p-m+1) - p^m(k-1)\cdots(k-m+1)}{p^{m-1}(p-1)\cdots(p-m+1)}\,\frac{k}{p}.$$

After elementary calculations we obtain

$$\phi_m(k,p) =$$

$$\frac{S_1 p^{m-2}k^{m-2}(p-k) - S_2 p^{m-3}k^{m-3}(p^2-k^2) + - \cdots + (-1)^{m-2}S_{m-1}(p^{m-1}-k^{m-1})}{p^{m-1}(p-1)\cdots(p-m+1)}\,\frac{k}{p}$$

$$\leqq \frac{S_1 p^{2m-3} + S_3 p^{2m-5} + \cdots}{p^{m-1}(p-1)\cdots(p-m+1)}.$$

(Here $S_1 = \sum_1^{m-1} k$, S_2 is the sum of the products of the numbers $1, 2, \cdots, k$ in pairs, etc.) The numerator of the last expression is a polynomial of degree $2m-3$ in p, and the denominator is a polynomial of degree $2m-2$. Therefore $\lim_{p\to\infty} \phi(k,p) = 0$. Thus

$$\lim_p \sum_{k=0}^{p} \binom{p}{k} \mu_{k,p-k} \left(\frac{k}{p}\right)^m = \mu_m.$$

By (9), $\mu_m = \int_0^1 t^m dH(t)$, and the theorem is proved.

We have the following consequences of Hausdorff's two theorems:

COROLLARY 1. *A moment sequence* $(\mu_n)_0^\infty$ *can be defined as one for which* (4) *or* (8) *is satisfied.*

COROLLARY 2. *Every moment sequence can be uniquely represented in the form*

$$\mu_n = \bar{\alpha}_n - \bar{\beta}_n = \int_0^1 t^n d\bar{g}_1(t) - \int_0^1 t^n d\bar{g}_2(t),$$

where $\bar{g}_1(t)$ *and* $\bar{g}_2(t)$ *are minimal components for* $H(t)$, *i.e. no monotonic function, except the zero function, can be subtracted from both of them.* $\mathrm{Var}_{[0,1]} H(t) = \bar{\alpha}_0 + \bar{\beta}_0$ *is a minimum.*

We conclude this section with some examples.

EXAMPLE 5. A sequence $(\mu_n)_0^\infty$ with alternating signs cannot be a moment sequence. Let $\mu_0 > 0$, $\mu_1 < 0$, etc. We have

$$|\mu_{0,n}| = \left| \mu_0 - \binom{n}{1} \mu_1 + - \cdots + (-1)^n \mu_n \right|$$

$$= \mu_0 + \binom{n}{1} |\mu_1| + \cdots > n|\mu_1|.$$

Condition (8) is not satisfied.

Likewise $\mu_0, \mu_1, \cdots, \mu_k, 0_{k+1}, \cdots, 0_{k+p}, \cdots$, where $k \geqq 1$ and $\mu_k \neq 0$, is not a sequence of moments; in fact, $\mu_{k,n} = \mu_k$ for any $n \geqq 0$, and (8) fails.

EXAMPLE 6. $(\mu_n)_0^\infty = (\rho^n)_0^\infty$.

1. For $0 \leqq \rho \leqq 1$ this sequence is absolutely monotonic, since $\mu_{m,n} = \rho^m (1-\rho)^n \geqq 0$. We obtain its integral representation by constructing the simplest step function, $g_\rho(t)$, with a single jump of amount $\delta = 1$ at the point $t = \rho$. Then $\rho^n = \int_0^1 t^n dg_\rho(t)$.

2. For $\rho > 1$ we have $\mu_n \to \infty$, and (8) is not satisfied.

3. For $\rho < 0$ we have $\mu_{0,n} = (1-\rho)^n \to \infty$, and again (8) is not satisfied.

EXAMPLE 7.

$$\mu_n = 1 \Big/ \binom{n+\alpha}{n}$$

is a moment sequence for each $\alpha > 0$.

In fact, it is well known that the Euler integrals

$$B(p,q) = \int_0^1 t^{p-1}(1-t)^{q-1} dt \qquad (p > 0;\ q > 0);$$

$$\Gamma(p) = \int_0^\infty t^{p-1}e^{-t}dt \qquad (p > 0)$$

are connected by

$$B(p,q) = \frac{\Gamma(p)\cdot\Gamma(q)}{\Gamma(p+q)}.$$

Then

$$\mu_n = \frac{\alpha n!\,\Gamma(\alpha)}{\Gamma(n+\alpha+1)} = \alpha B(n+1,\alpha).$$

Thus

$$\mu_n = \int_0^1 t^n d[1-(1-t)^\alpha].$$

We note also that the sequence $\{1/(n+1)!\}_0^\infty$ is not a moment sequence, but since the proof is rather difficult [**3**] we shall not go into it.

§2. The linear functional determined by a moment sequence

The set of functions $\{f(x)\}$, each continuous on $[0,1]$, becomes the complete metric space C if we suppose the distance between two elements $f_1(x)$ and $f_2(x)$ of the set to be defined by the nonnegative number $\max_{[0,1]}|f_1(x) - f_2(x)|$.

We say that a sequence $f_1(x), f_2(x), \cdots, f_n(x), \cdots$ converges in C if $\max_{[0,1]}|f_{n+p}(x) - f_n(x)| \to 0$. Thus convergence in C is defined as uniform convergence of $f_n(x)$ on $0 \leqq x \leqq 1$, and consequently C always contains a limit element $f(x)$ of the sequence $f_n(x)$ [**2**].

Let a functional $F(f)$ be defined on C. This means that to each $f(x) \in C$ there is assigned a unique real number $F(f)$.

The functional $F(f)$ is called *linear* if F satisfies the following conditions:

1) $F(f_1 + f_2) = F(f_1) + F(f_2)$ (additivity);

2) $F(af) = a \cdot F(f)$ for each real number (homogeneity); in particular, $F(0) = 0$.

3) $\lim_{n\to\infty} F(f_n) = F(f)$ if $f_n(x)$ converges to $f(x)$ in the sense defined above (continuity).

We show now that the last condition is equivalent to a different one.

THEOREM. *A necessary and sufficient condition that an additive homogenous functional is continuous is that $F(f)$ is bounded, i.e.*

$$|F(f)| \leqq K \max_{[0,1]} |f(x)| \quad \textit{for all } f(x) \in C. \tag{10}$$

1. Let the linear functional $F(f)$ be unbounded. Then there is a

sequence $\{f_n(x)\}_{n=1}^{\infty}$ such that $|F(f_n)| > n \cdot \max_{[0,1]}|f_n(x)|$. Form the sequence $\{\phi_n(x)\}$, where

$$\phi_n(x) = \frac{f_n(x)}{n \cdot \max\limits_{[0,1]}|f_n(x)|};$$

then $\phi_n(x) \to 0$ as $n \to \infty$. Furthermore,

$$|F(\phi_n)| = \frac{1}{n \max\limits_{[0,1]}|f_n(x)|} \cdot |F(f_n)| > 1,$$

but the property of continuity requires that $F(\phi_n) \to F(0) = 0$. This establishes the necessity.

2. Let the functional F be bounded on C: $|F(f)| \leqq K \max|f(x)|$. Suppose $f_n(x) \to f(x)$ in C, i.e. $\max|f_n(x) - f(x)| \to 0$ as $n \to \infty$. Then $|F(f_n) - F(f)| = |F(f_n - f)| \leqq K \max|f_n(x) - f(x)|$. Hence $\lim_{n\to\infty} F(f_n) = F(f)$ and the functional is continuous. This establishes the sufficiency.

The smallest possible number K in (10) is called the norm of the linear functional F and is denoted by N. Thus if $\max|f(x)| = 1$, i.e. if $f(x)$ is a reduced function, we have

$$N = \operatorname{Sup}|F(f)|. \tag{11}$$

For an example of a linear functional take any $H(t)$ of bounded variation on $[0,1]$ and define

$$F(f) = \int_0^1 f(t)\,dH(t).$$

The additivity and homogeneity of this functional follow from the simplest properties of the integral, and the continuity of F follows from its boundedness (see (2)):

$$|F(f)| \leqq \max_{[0,1]}|f(t)| \cdot \operatorname*{Var}_{[0,1]} H(t).$$

Note that any such functional, evaluated on the sequence of powers $1, x, x^2, \cdots, x^n, \cdots$ generates a moment sequence:

$$F(x^n) = \int_0^1 t^n dH(t) = \mu_n.$$

An important fact, established by F. Riesz [**2**], is that the converse statement holds.

RIESZ'S THEOREM. *Every linear functional on the space C can be represented in the form*

$$F(f) = \int_0^1 f(t)\,dH(t), \tag{12}$$

where $H(t)$ *is a function of bounded variation on* $[0,1]$.

This can easily be proved by using Hausdorff's Second Theorem. Let $F(f)$ be a given linear functional on C.

We shall show that the values of the functional on $(x^n)_{n=0}^{\infty}$, i.e. $F(1), \cdots, F(x^n), \cdots$ form a moment sequence. In fact, putting $F(x^n) = \mu_n$, we have

$$\mu_{m,n} = \mu_m - \binom{n}{1}\mu_{m+1} + \cdots + (-1)^n \mu_{m+n};$$

consequently

$$\mu_{m,n} = F\left[x^m - \binom{n}{1} x^{m+1} + \cdots + (-1)^n x^{m+n}\right] = F[x^m(1-x)^n].$$

Putting $\epsilon_{m,n} = \operatorname{Sgn}\mu_{m,n}$, we have, by the boundedness of the functional,

$$\left|\sum_{k=0}^{p}\binom{p}{k}\mu_{k,p-k}\epsilon_{k,p-k}\right| = \left|F\left[\sum\binom{p}{k}x^k(1-x)^{p-k}\epsilon_{k,p-k}\right]\right|$$

$$\leqq K\cdot\max\left|\sum\binom{p}{k}x^k(1-x)^{p-k}\epsilon_{k,p-k}\right| \leqq K;$$

condition (4) is satisfied, and $(\mu_n)_0^\infty$ is a moment sequence. Consequently there is an $H(t)$ such that $F(x^n) = \int_0^1 t^n dH(t)$, and hence (12) holds for an arbitrary polynomial $f(t)$. The possibility of extending F to the whole space C, and indeed in just one way, follows from the classical theorem of Weierstrass [2], which states that every continuous function on a finite interval can be represented as the limit of a sequence of polynomials that converges uniformly on the interval:

$$f(x) = \lim_{n\to\infty} P_n(x).$$

It follows from this theorem that by the continuity of the functional

$$F(f) = \lim_{n\to\infty} F(P_n) = \lim_{n\to\infty}\int_0^1 P_n(t)\,dH(t) = \int_0^1 f(t)\,dH(t),$$

and this completes the proof.

Thus every moment sequence $(\mu_n)_0^\infty$ on $[0,1]$, and only such a sequence, determines a linear functional on C; consequently we may use the expression "Given a linear functional $(\mu_n)_0^\infty$ on C."

We note that the extension of such a functional to the whole space C can always be carried out by using the Bernšteĭn polynomials [**2**]: with every continuous function $f(x)$ on $[0,1]$ there is associated its Bernšteĭn polynomial:

$$B_n(x,f) = \sum_0^n \binom{n}{k} x^k(1-x)^{n-k} f\left(\frac{k}{n}\right).$$

We have the formula $\lim_{n\to\infty} B_n(x,f) = f(x)$, uniformly in $[0,1]$.

From this it follows that

$$F(f) = \lim_{n\to\infty} G(B_n) = \lim_{n\to\infty} \sum_{k=0}^n \binom{n}{k} \mu_{k,n-k} f\left(\frac{k}{n}\right). \tag{13}$$

Let $F(f)$ be defined by the sequence $(\mu_n)_0^\infty$, where $\mu_n = \int_0^1 t^n dH(t)$; let $\mu_n = \bar{\alpha}_n - \bar{\beta}_n$ and $H(t) = \bar{g}_1(t) - \bar{g}_2(t)$, where $(\bar{\alpha}_n)$ and $(\bar{\beta}_n)$ are minimal components for μ_n, and $g_1(t)$, $g_2(t)$ are minimal components for $H(t)$.

Then we have the following expression for the norm N of the functional:

$$\begin{aligned} N &= \lim_{p\to\infty} \sum^p \binom{p}{k} |\mu_{k,p-k}| = \bar{\alpha}_0 + \bar{\beta}_0 = \operatorname*{Var}_{[0,1]} H(t) \\ &= \operatorname*{Var}_{[0,1]} \bar{g}_1(t) + \operatorname*{Var}_{[0,1]} \bar{g}_2(t). \end{aligned} \tag{14}$$

In fact, for any $f(x)$ we obtain from (13)

$$|F(f)| \leq \lim_p \sum_0^p \binom{p}{k} |\mu_{k,p-k}| = \lim M_p;$$

but

$$M_p = F\left[\sum_0^p \binom{p}{k} x^k(1-x)^{p-k}\epsilon_{k,p-k}\right] \leq N,$$

and hence $\lim M_p = N$.

It follows from (7) that $N = \bar{\alpha}_0 + \bar{\beta}_0$, and the rest of (14) follows from Hausdorff's Second Theorem. In the special case when the functional is defined by an absolutely monotonic sequence $(\alpha_n)_0^\infty$ we have $N = \alpha_0$.

We now introduce the concept of an extremal function $\phi(x)$ of a linear functional F defined on C.

If $\phi(x) \in C$ and $\max_{[0,1]}|\phi(x)| = 1$ (reduced function), we call $\phi(x)$ *extremal* if F attains its norm on $\phi(x)$, i.e. if $F(\phi) = +N$. If $\phi(x)$ is a polynomial, we call it an *extremal polynomial*.

In the special case when the defining sequence $(\alpha_n)_0^\infty$ is absolutely monotonic, the function $\phi(x) = 1$ is extremal, since $F(\phi) = \alpha_0 = N$.

In the general case there does not have to be an extremal function in C. If $(\mu_n)_0^\infty$ defines F, and $(\bar{\alpha}_n)_0^\infty$ and $(\bar{\beta}_n)_0^\infty$ are its minimal components defining functionals F_1 and F_2, then if $\phi(x)$ is an extremal function we have $N = F(\phi) = \bar{\alpha}_0 + \bar{\beta}_0 = F_1(\phi) - F_2(\phi)$, whence necessarily $F_1(\phi) = \bar{\alpha}_0$ and $F_2(\phi) = -\bar{\beta}_0$. Therefore $\phi(x)$ must be extremal for F_1, and $-\phi(x)$ for F_2. Letting $H(t) = \bar{g}_1(t) - \bar{g}_2(t)$ be the integrator function of (μ_n) and $\bar{g}_1$ and $\bar{g}_2$ its minimal components, we find that at the points of increase of $\bar{g}_1$ and $\bar{g}_2$ the function ϕ must be equal to $+1$ or -1 (Example 9).

EXAMPLE 8. $(\mu_n)_0^\infty = (1/(n+1))_0^\infty$. Then $F(f) = \int_0^1 f(t)\,dt$. The norm N is 1 and is attained only for $\phi(x) = 1$, since $\int_0^1 f(t)\,dt < 1$ for all $f(t)$ other than unity.

EXAMPLE 9. $\mu_n = n/(n+1)(n+2)$. Here $F(f) = \int_0^1 f(t)\,d(t^2 - t)$;

$$\operatorname*{Var}_{[0,1]}(t^2 - t) = \frac{1}{2}.$$

The minimal components of $H(t) = t^2 - t$ are

$$\bar{g}_1(t) = \begin{cases} 0 \text{ on } \left[0, \dfrac{1}{2}\right] \\ t^2 - t + \dfrac{1}{4} \text{ on } \left[\dfrac{1}{2}, 1\right] \end{cases};$$

$$\bar{g}_2(t) = \begin{cases} -t^2 + t \text{ on } \left[0, \dfrac{1}{2}\right] \\ \dfrac{1}{4} \text{ on } \left[\dfrac{1}{2}, 1\right] \end{cases}.$$

The norm is $N = \frac{1}{2}$. There is no extremal function $\phi(t)$, since we should have to have both $\phi(\frac{1}{2}) = 1$ and $\phi(\frac{1}{2}) = -1$.

EXAMPLE 10. $\mu_n = (1 - 3^n)/2^{2n}$ $(n = 0, 1, \cdots)$. Here $\bar{\alpha}_n = (\frac{1}{4})^n$ and $\bar{\beta}_n = (\frac{3}{4})^n$; $N = 2$ (Example 6).

An extremal function is $\phi(x) = \sin 2\pi x$, since because $H(t)$ is a step function (Example 6), we have

$$F(\phi) = \int_0^1 \phi(t)\,dH(t) = \phi\left(\frac{1}{4}\right) - \phi\left(\frac{3}{4}\right) = 2;$$

there is also an extremal polynomial $T_3(x) = 32x^3 - 48x^2 + 18x - 1$, since $F(T_3) = T_3(\frac{1}{4}) - T_3(\frac{3}{4}) = 2$; $\max_{[0,1]} |T_3(x)| = 1$.

Part One

EXTREMAL POLYNOMIALS OF FINITE FUNCTIONALS

CHAPTER I

SEGMENT-FUNCTIONALS ON POLYNOMIALS

§1. Extremal polynomials of a moment sequence

A linear functional F on C, or (what is the same thing) a moment sequence $(\mu_n)_0^\infty$, does not in general have an extremal function. Our immediate problem is to find conditions under which not only is there an extremal function $\phi(x)$, but $\phi(x)$ is a polynomial.

We begin with absolutely monotonic sequences $(\alpha_n)_0^\infty$, i.e. those for which all $\alpha_{n,k} \geqq 0$. We do not have to consider the sequence $(-\alpha_n)_0^\infty$ since all its properties are immediately obtainable from those of $(\alpha_n)_0^\infty$.

We shall use the term "extremal polynomial of a functional" only for an extremal polynomial $Q_n(x)$ of positive degree; thus we exclude $\phi(x) \equiv 1$ (Example 8).

We shall denote by $Q_n(\bar{\mu})$ the value $F(Q_{\underline{n}})$, and by $Q_{\underline{n}}(x)$ a polynomial of degree precisely n.

THEOREM 1 [5]. *A necessary and sufficient condition that the absolutely monotonic sequence*

$$\alpha_0, \alpha_1, \cdots, \alpha_p, \cdots; \quad (\alpha_0 > 0) \tag{15}$$

has an extremal polynomial $P(x) \not\equiv 1$ is that the integrator function $g(t)$ of (15) is a step function with a finite number of jumps on $0 \leqq t \leqq 1$; and if $\sigma_1, \sigma_2, \cdots, \sigma_s$ are the abscissas of the points of increase of $g(t)$ and $\delta_1, \delta_2, \cdots, \delta_s$ are the corresponding (positive!) jumps, then every reduced polynomial $P(x)$ for which $P(\sigma_k) = +1$ $(k = 1, 2, \cdots, s)$ is extremal.

NECESSITY. Put $P(\bar{\alpha}) = \alpha_0$ $(= N)$. There is certainly a point x_0 at which $P(x_0) < 1$, with $0 < x_0 < 1$. Then there is an interval $[\alpha, \beta]$, where $0 \leqq \alpha \leqq x_0 \leqq \beta \leqq 1$, in which $P(x) < 1$; applying the mean value theorem, we obtain

$$\int_\alpha^\beta P(t)dg(t) = P(\xi)\cdot \operatorname*{Var}_{[\alpha,\beta]} g(t) < \operatorname*{Var}_{[\alpha,\beta]} g(t),$$

but

$$\alpha_0 = \int_0^1 P(t)\,dg(t) = \int_0^\alpha + \int_\alpha^\beta + \int_\beta^1$$

$$= P(\xi_1)\cdot \operatorname*{Var}_{[0,\alpha]} g(t) + P(\xi_2)\operatorname*{Var}_{[\alpha,\beta]} g(t) + P(\xi_3)\operatorname*{Var}_{[\beta,1]} g(t).$$

The last equation can hold only if $\operatorname{Var} g(t) = 0$, i.e. $g(t) = \text{const.}$ on $[\alpha,\beta]$. Thus $g(t)$ is a step function, and can have discontinuities only at points where $P(x) = 1$, i.e. at finitely many points.

SUFFICIENCY. Let $g(t)$ be the integrator function for the sequence (15), let it be a step function on $[0,1]$, and let $(\sigma_i)_1^s$ be the abscissas of its points of discontinuity. If $0 < \sigma_i < 1$ then $g(\sigma_i+) - g(\sigma_i-) = \delta_i > 0$; if $\sigma_1 = 0$ then $g(0) = 0$ and $g(0+) = \delta_1 > 0$; if $\sigma_s = 1$ then $g(1) - g(1-) = \delta_s > 0$. We have

$$\alpha_0 = \int_0^1 dg(t) = \sum_1^s \delta_i, \qquad \alpha_k = \int_0^1 t^k dg(t) = \sum_{i=1}^s \delta_i \sigma_i^k, \cdots.$$

If $P_n(x) = \sum_0^n p_k x^k$ is a reduced polynomial then

$$P_n(\bar\alpha) = \int_0^1 \sum_0^n p_k t^k dg(t) = \sum_{i=1}^s P_n(\sigma_i)\cdot\delta_i.$$

Hence it is clear that a necessary and sufficient condition for $P_n(\bar\alpha) = \alpha_0 = \sum_1^s \delta_i$ is that $P_n(\sigma_i) = +1$. This establishes the sufficiency.

COROLLARY. *If $P_{\underline{n}}(x)$ is an extremal polynomial of lowest degree $(n > 0)$ for the functional F_α (15), its general form can be written immediately for each of the four possible distributions of the discontinuities $(\sigma_i)_1^s$ on $[0,1]$. For*

I $\quad 0 < \sigma_1 < \cdots < \sigma_s < 1, \quad P_{\underline{n}}(x) = 1 - C\prod_1^s (x-\sigma_i)^2;$

II $\quad 0 = \sigma_1 < \cdots < \sigma_s < 1, \quad P_{\underline{n}}(x) = 1 - Cx\prod_2^s (x-\sigma_i)^2;$

III $\quad 0 < \sigma_1 < \cdots < \sigma_s = 1, \quad P_{\underline{n}}(x) = 1 - C(1-x)\prod_1^{s-1} (x-\sigma_i)^2;$

IV $\quad 0 = \sigma_1 < \cdots < \sigma_s = 1, \quad P_{\underline{n}}(x) = 1 - Cx(1-x)\prod_2^{s-1} (x-\sigma_i)^2.$

Here C is a constant and we must have $0 \leqq 1 - P_{\underline{n}}(x) \leqq 2$ *with* $C > 0$.

Clearly the degree cannot be decreased, since every point where $P_{\underline{n}}(x) = +1$ is a discontinuity of $g(t)$.

From now on we shall call the polynomial $R_s(x) = \prod_1^s (x - \sigma_i)$ the *resolvent* of the extremal polynomial $P_n(x)$ of smallest degree, and the factor multiplying C in the formulas I—IV will be called (by convention) the squared resolvent.

Thus if $g(t)$ has only interior points of discontinuity, $P_{\underline{n}}(x)$ is of even degree $n = 2s$, with leading coefficient $p_n < 0$; in the second case $n = 2s - 1$ and $p_n < 0$; in the third, $n = 2s - 1$ and $p_n > 0$; in the fourth, $n = 2s - 2$ and $p_n > 0$.

Thus the extremal polynomial of lowest degree uniquely reflects the four possibilities.

If $P_{\underline{n}}(x)$ is known and $(\alpha_k)_0^n$ is given, it is easy to construct $g(t)$ by finding the maximum points $\sigma_1, \sigma_2, \cdots, \sigma_n$ of $P_{\underline{n}}(x)$ on $[0, 1]$ and solving the system $\sum_{i=1}^s \sigma_i^k \delta_i = \alpha_k$ $(k = 0, 1, \cdots, s - 1)$ for the jumps δ_i of $g(t)$. In each case $s \leqq \frac{1}{2}n + 1$.

We now consider the general case: $\mu_p = \int_0^1 t^p dH(t)$, where $H(t)$ is of bounded variation.

THEOREM 2. *The sequence of moments* $(\mu_i)_0^\infty$ *has an extremal polynomial* $Q_n(x) \not\equiv \pm 1$ *if and only if its minimal components* $(\bar{\alpha}_i)_0^\infty$ *and* $(\bar{\beta}_i)_0^\infty$ *both have extremal polynomials.*

NECESSITY. Let the norm of the functional defined by $(\mu_k)_0^n$ be N and let the sequence have an extremal polynomial $Q(x)$; we have $Q(\bar{\mu}) = +N$, and then $N = \alpha_0 + \beta_0 = Q(\bar{\alpha}) - Q(\bar{\beta}) \leqq |Q(\bar{\alpha})| + |Q(\bar{\beta})| \leqq \alpha_0 + \beta_0$, i.e. $Q(\bar{\alpha}) = \alpha_0$ and $Q(\bar{\beta}) = -\beta_0$.

Consequently $Q(x)$ is extremal for $(\alpha_i)_0^\infty$ and $-Q(x)$ is extremal for $(\beta_i)_0^\infty$. By Theorem 1, $(\alpha_i)_0^\infty$ and $(\beta_i)_0^\infty$ have step-function integrators $g_1(t)$ and $g_2(t)$.

SUFFICIENCY. Suppose that $Q_{p_1}(x)$ and $Q_{p_2}(x)$ are reduced polynomials such that $Q_{p_1}(\bar{\alpha}) = \alpha_0$ and $Q_{p_2}(\bar{\beta}) = \beta_0$ $(p_1 > 0,\ p_2 > 0)$, and suppose that both polynomials are of the lowest possible degree. This means that $Q_{p_1}(x)$ takes the value $+1$ at all points (a_i) of discontinuity of $g_1(t)$, and that $Q_{p_2}(x)$ takes the value $+1$ at all points (b_i) of discontinuity of $g_2(t)$. We now show how to construct an extremal polynomial $Q(x)$ for $(\mu_i)_0^\infty$.

We note first that $(\alpha_i)_0^\infty$ and $(\beta_i)_0^\infty$ are minimal components if and only if the monotonic step functions $g_1(t)$ and $g_2(t)$ have no common points of discontinuity. In fact, let the points of discontinuity of the two functions be $(a_i)_1^{m_1}$ and $(b_i)_1^{m_2}$. Suppose that $a_{k_1} = b_{k_2} = c$ with corresponding jumps δ_a and δ_b. Construct an absolutely monotonic sequence $(\gamma_i)_0^\infty$ for which the integrator $g_3(t)$ is a step function with the single jump $\delta_c > 0$ at the point c, and let δ_c be the smaller of δ_a and δ_b. Then both differences $g_1(t) - g_3(t)$ and $g_2(t) - g_3(t)$ are nondecreasing. Let $\gamma_i = \int_0^1 t^i dg_3(t)$; then $(\alpha_i - \gamma_i)_0^\infty = (\alpha_i')_0^\infty$ and $(\beta_i - \gamma_i)_0^\infty = (\beta_i')_0^\infty$ are two other components for $(\mu_i)_0^\infty = (\alpha_i' - \beta_i')_0^\infty$; now we have $\alpha_0' + \beta_0' = N - 2\gamma_0 < N$, which is impossible.

Suppose that $g_1(t)$ and $g_2(t)$ have no common points of discontinuity; then, since they are step functions, there is no nondecreasing function ($\not\equiv 0$) that can be subtracted from both without making one or the other fail to be monotonic.

We now construct $Q(x)$ for $(\mu_i)_0^\infty$ [7]. By the corollary of Theorem 1, every extremal polynomial of lowest degree for each of the minimal components has one of the forms I, II, III, IV. (The case $C = 0$ is, by hypothesis, discarded as trivial.) Then every extremal polynomial of higher degree can be expressed in the form $1 - R^2(x) \cdot \phi(x)$, where $\phi(x) \geqq 0$ is any polynomial for which $0 \leqq R^2(x)\phi(x) \leqq 2$ for $0 \leqq x \leqq 1$. Hence if an absolutely monotonic sequence has an extremal polynomial of degree precisely n, it also has extremal polynomials of every higher degree.

Thus in the general case the required $Q(x)$ must simultaneously satisfy

$$Q(x) = 1 - R_1^2(x)\phi(x) \equiv -1 + R_2^2(x)\psi(x), \tag{16}$$

i.e.

$$R_1^2(x)\phi(x) + R_2^2(x)\psi(x) \equiv 2. \tag{17}$$

Here $R_1^2(x)$ and $R_2^2(x)$ are the "squared resolvents" of $Q_{p_1}(x)$ and $Q_{p_2}(x)$, supposed known, and $\phi(x)$ and $\psi(x)$ are to be found, with

$$\phi(x) \geqq 0 \quad \text{and} \quad \psi(x) \geqq 0; \quad 0 \leqq x \leqq 1.$$

By the Euclidean algorithm, ϕ and ψ can be constructed by using (17), and in just one way since $R_1^2(x)$ and $R_2^2(x)$ are relatively prime. The degrees of $\phi(x)$ and $\psi(x)$ cannot exceed $n_2 - 1$ and $n_1 - 1$ if n_1 and n_2 are the degrees of $R_1^2(x)$ and $R_2^2(x)$.

If $\phi(x)$ and $\psi(x)$ turn out to be nonnegative on $[0,1]$, then $Q(x)$ (of the lowest possible degree) has been found, and satisfies (16).

If $\phi(x)$ and $\psi(x)$ are not both nonnegative, then (17) implies that it is possible to construct a polynomial $\lambda(x)$ that satisfies

$$R_1^2(x)\,[\phi(x) + R_2^2(x)\lambda(x)] + R_2^2(x)\,[\psi(x) - R_1^2\;(x)\lambda(x)] \equiv 2,$$

and makes $\phi(x) + R_2^2(x)\lambda(x)$ and $\psi(x) - R_1^2(x)\lambda(x)$ nonnegative on $[0,1]$. In fact, we have

$$-\frac{\phi(x)}{R_2^2(x)} \leqq \lambda(x) \leqq \frac{\psi(x)}{R_1^2(x)}. \tag{18}$$

The strip in which $\lambda(x)$ must be chosen has width

$$\frac{2}{R_1^2(x)\,R_2^2(x)} > 0.$$

If we take an arbitrary continuous curve inside this strip, then by Weierstrass's theorem we can approximate it by a polynomial $\lambda(x)$ of sufficiently high degree which also lies inside the strip.

After choosing $\lambda(x)$ we can write

$$\begin{aligned} Q(x) &= 1 - R_1^2(x)\,[\phi(x) + \lambda(x)\,R_2^2(x)] \\ &= -1 + R_2^2(x)\,[\psi(x) - \lambda(x)\,R_1^2(x)]. \end{aligned} \tag{19}$$

Since this polynomial is reduced, i.e. $\max|Q(x)| = 1$, Theorem 2 is established.

The strip defined by (18) will be called the *reducibility strip* for $\lambda(x)$.

COROLLARY. *If the moment sequence* $(\mu_k)_0^\infty$ *has an extremal polynomial* $Q_n(x)$, *the numbers* μ_k $(k = 0, 1, 2, \cdots)$ *are algebraic sums of terms selected from a finite number of geometric progressions:*

$$\mu_k = \sum_{i=1}^{s} \delta_i \sigma_i^k, \quad \textit{where} \quad 0 \leqq \sigma_i \leqq 1 \quad \textit{and} \quad Q_n(\sigma_i) = \operatorname{Sgn} \delta_i;$$

$(\sigma_i)_1^s$ *are the points of discontinuity of the integrator function* $H(t)$. *The converse is also true.*

We say that such a sequence is of *nodal structure*, and call the points σ_i the nodes of $(\mu_i)_0^\infty$. The norm N of the corresponding functional F_μ is

$$N = Q_n(\bar{\mu}) = \sum_1^s |\delta_i|.$$

We note some special cases of nodal structure.

1. If $\sigma_s = 1$ and $\delta_s > 0$, then $\lim_{p\to\infty} \mu_p = \lim_{p\to\infty} \alpha_p = \delta_s$, i.e. $\lim_{p\to\infty} \beta_p = 0$; if $\delta_s < 0$, then $\lim \mu_p = -\lim \beta_p = \delta_s$ and $\lim \alpha_p = 0$.

2. If $\sigma_1 = 0$, then

$$\delta_1 = \lim_{p\to\infty} \mu_{0,p}\left[\text{where } \mu_{0,p} = \mu_0 - \binom{p}{1}\mu_1 + \cdots + (-1)^p\mu_p\right].$$

In fact, if $\mu_p = \int_0^1 t^p dH(t)$, then

$$\mu_{0,p} = \int_0^1 (1-t)^p dH(t) = \int_0^1 \tau^p d[H(1) - H(1-\tau)] = \int_0^1 \tau^p dh(\tau).$$

Consequently δ_1 with $t = 0$ for $H(t)$ becomes the jump δ_1 with $\tau = 1$ for $h(\tau)$. In general, $h(\tau)$ has discontinuities $1 - \sigma_i$ with jumps δ_i, and if $Q_n(x)$ is an extremal polynomial for $(\mu_p)_0^\infty$ then $Q_n(1-x)$ is an extremal polynomial for $(\mu_{0,p})_0^\infty$.

THEOREM 3. *No sequence $(\mu_i)_0^\infty$ of nodal structure $\mu_n = \sum_{i=1}^s \Delta_i \rho_i^n$ $(n = 0, 1, 2, \cdots)$ for which some node ρ is outside $[0,1]$ can be a sequence of moments.*

1. Let ρ_1 be the node of largest absolute value and let $|\rho_1| > 1$; then

$$|\mu_n| = |\rho_1^n| \cdot \left|\Delta_1 + \sum_2^s \Delta_i \left(\frac{\rho_i}{\rho_1}\right)^n\right| = |\rho_1|^n A_n;$$

here $\lim_{n\to\infty} A_n = |\Delta_1| > 0$ and $\lim_{n\to\infty} |\rho_1|^n = \infty$; consequently $(\mu_i)_0^\infty$ is not a moment sequence.

2. Let $-1 \leqq \rho_i \leqq 1$ for all nodes. Let $(\rho_i')_1^{s_1}$ be the nodes in $[0,1]$ and let their weights be $(\Delta_i')_1^{s_1}$; let $(\rho_i'')_1^{s-s_1}$ be the remaining nodes, with weights $(\Delta_i'')_1^{s-s_1}$. Then if we put $\nu_n = \sum_{i=1}^{s_1} \Delta_i' \rho_i'^n$ and $\lambda_n = \sum_{i=1}^{s-s_1} \Delta_i'' \rho_i''^n$, we have $\mu_n = \lambda_n + \nu_n$; clearly $(\nu_n)_0^\infty$ is a moment sequence. It is easily verified that a sequence of the form $(\Delta \cdot q^n)_0^\infty = (\vartheta_n)_0^\infty$ with $-1 \leqq q < 0$ is not a moment sequence, since $\vartheta_{0,n} = \Delta \cdot (1-q)^n \to \infty$ as $n \to \infty$. Since $(\nu + \lambda)_{0,n} = \nu_{0,n} + \lambda_{0,n}$ and since $(\lambda_{0,n})_0^\infty$ satisfies the hypotheses of case 1, we have $\lambda_{0,n} \to \infty$. This completes the proof.

THEOREM 4. *A necessary and sufficient condition for a given sequence $(\mu_p)_0^\infty$ to be a moment sequence and have an extremal polynomial is that the following limits exist and satisfy the specified conditions:*

1. $\lim_{p\to\infty} \mu_p = \delta_1$;
2. $\lim |\nu_p|^{1/p} = \rho_2 > 0$, *where* $\nu_p = \mu_p - \delta_1$;
3. $\lim(\nu_p/\rho_2^p) = \delta_2\ (\neq 0)$;
4. $\lim |\lambda_p|^{1/p} = \rho_3 > 0$, *where* $\nu_p - \delta_2\rho_2^p = \lambda_p$;
5. $\lim(\lambda_p/\rho_3^p) = \delta_3\ (\neq 0)$;

and so on; also that after a finite number of steps, say $s-1$, we obtain a sequence $(\xi_p)_0^\infty$ such that $\xi_p = 0$ for $p \geqq 1$ and $\xi_0 = \delta_s$.

In this case the integrator function $H(t)$ for $(\mu_p)_0^\infty$ is a step function with discontinuities at $1 = \rho_1 > \rho_2 > \cdots > \rho_{s-1} > \rho_s = 0$, and the jumps are $\delta_1, \delta_2, \cdots, \delta_s$.

NECESSITY. If $(\mu_p)_0^\infty$ is a moment sequence with an extremal polynomial, then (Theorem 2) its integrator $H(t)$ is a step function with a finite number of discontinuities, say at $(\rho_i)_1^s$, where $1 \geqq \rho_1 > \rho_2 > \cdots > \rho_s \geqq 0$ with jumps $\delta_1, \delta_2, \cdots, \delta_s$. Then $\mu_p = \sum_{i=1}^s \rho_i^p \delta_i$; if $\rho_1 = 1$, i.e. $\delta_1 \neq 0$, then

$$\delta_1 = \lim_{p \to \infty} \mu_p; \quad \nu_p = \sum_{i=2}^s \rho_i^p \delta_i; \quad \frac{\nu_p}{\rho_2^p} = \sum_{i=3}^s \left(\frac{\rho_i}{\rho_2}\right)^p \delta_i + \delta_2;$$

consequently $\delta_2 = \lim_p \nu_p / \rho_2^p \neq 0$ and the ν_p, from a certain index on, are of the same sign; in fact $\operatorname{Sgn} \nu_p = \operatorname{Sgn} \delta_2$. Therefore $|\nu_p|^{1/p}/\rho_2 = (|\delta_2| + \epsilon_p)^{1/p} \to 1$ as $p \to \infty$; then $\rho_2 = \lim |\nu_p|^{1/p} < 1$. Starting over with ρ_2 and then δ_2 we obtain $(\lambda_p)_0^\infty$, where $\lambda_p = \nu_p - \delta_2 \rho_2^p$; and then we repeat the process. After a finite number of steps we arrive at $(\xi_p)_0^\infty$ in which either all $\xi_p = 0$, in which case $\delta_s = 0$, or $\xi_p = 0$ with $p \geqq 1$ and $\xi_0 \neq 0$, and then $\xi_0 = \delta_s$.

SUFFICIENCY. If the hypotheses of the theorem are satisfied, then $(\mu_p)_0^\infty$ evidently is of nodal structure,

$$\mu_p = \sum_{i=1}^s \delta_i \rho_i^p \quad (p = 0, 1, 2, \cdots), \quad \text{where } 0 \leqq \rho_i \leqq 1.$$

Consequently $(\mu_p)_0^\infty$, as the algebraic sum of a finite number of moment sequences (decreasing geometric progressions, Example 6) is itself a moment sequence. Its norm is $N = \sum_1^s |\delta_i|$.

However, we can always construct a polynomial $Q_n(x) = \sum_0^n q_k x^k$, with the following properties

$$\max_{[0,1]} |Q_n(x)| = 1; \quad Q_n(\rho_i) = \pm 1, \quad \text{and} \quad \operatorname{Sgn} Q_n(\rho_i) = \operatorname{Sgn} \delta_i$$

(its degree n may be very large: Theorem 2, sufficiency). But then

$$Q_n(\bar{\mu}) = \sum_0^n q_k \mu_k = \sum_{k=0}^n q_k \sum_{i=1}^s \delta_i \rho_i^k = \sum_{i=1}^s \delta_i Q_n(\rho_i) = \sum_1^n |\delta_i| = N,$$

and $Q_n(x)$ is an extremal polynomial.

We note some results that are fundamental for our further work and follow from the theorems we have developed [8].

Suppose that $(\mu_i)_0^n$ is a finite sequence of real numbers; we call it a "segment of numbers." We may suppose that it defines a linear functional F_n on the set $\{P_n(x)\}$ of polynomials, by putting $F_n(x^i) = \mu_i$ $(i = 0, 1, 2, \cdots)$; then we call the given segment a *segment-functional.* Since the space $\{P_n(x)\}$ is finite-dimensional [7], the norm N_n of this functional is attained for some reduced polynomial $Q_n(x)$, which is then an extremal polynomial, i.e. $Q_n(\bar{\mu}) = F_n(Q) = +N_n$. In addition, by the Hahn-Banach Theorem [9], this functional can be extended to the set of polynomials of degree $n+1$ in such a way that its norm is preserved (i.e. not increased); this is equivalent to saying that the given segment $(\mu_i)_0^n$ can be extended by one number μ_{n+1}. Forming $(\mu_i)_0^{n+1}$, we say that the original segment-functional has been "extended in the best possible way."

It is obvious that if $(\mu_i)_0^\infty$ is a moment sequence with an extremal polynomial $Q_{\underline{n}}(x)$, then its segment-functionals $(\mu_i)_0^n$ have the same norm N_n as the whole sequence: $N_n = N$, and similarly the segments $(\mu_i)_0^{n+1}, (\mu_i)_0^{n+2}, \cdots$ have the same norm $N_{n+p} = N$ $(p = 1, 2, \cdots)$. Thus $(\mu_i)_0^\infty$ can be obtained from $(\mu_i)_0^n$ by best extensions of $(\mu_i)_0^n$ by one term, then by two, etc.

In the later sections of this chapter we shall be concerned with clarifying the character of such extensions and with their construction; we shall need to investigate the norm as a function of one of the parameters μ_k. Here we establish that $N(\mu_k)$ is a continuous function.

THEOREM 5. *For any segment $(\mu_i)_0^n$, if one of the parameters μ_k is varied $(k = 0, 1, \cdots, n)$, the norm $N(\mu_k)$ of the segment-functional is a continuous function of μ_k.*

In fact, we have

$$N(\mu_k) - \max|p_k| \cdot |h| \leqq N(\mu_k + h) \leqq N(\mu_k) + \max|p_k| \cdot |h|,$$

where p_k is the coefficient of x^k in an arbitrary reduced polynomial $P_n(x)$. Thus $|N(\mu_k + h) - N(\mu_k)| \leqq \max|p_k| \cdot |h|$; but the set $\{p_k\}$ of coefficients of reduced polynomials is bounded by the norm of the segment-functional $0_0, 0_1, \cdots, 0_{k-1}, 1_k, 0_{k+1}, \cdots, 0_n$. This establishes the theorem.

§2. Absolutely monotonic segments and their best extension [5]

DEFINITION. A segment $(\alpha_i)_0^n$ is called *absolutely monotonic* if there is at least one absolutely monotonic sequence of which (α_i) forms the first $n+1$ terms, i.e. if $\alpha_0, \alpha_1, \cdots, \alpha_n, \alpha_{n+1}, \cdots, \alpha_{n+p}, \cdots$ is absolutely monotonic; in the contrary case we say that the segment is *non-absolutely-*

monotonic. (Segments of the form $(-\alpha_i)_0^n$ with absolutely monotonic $(\alpha_i)_0^n$ can be left out of consideration.)

Since the norm of the segment-functional $(\alpha_i)_0^n$ is $N = \alpha_0$, it is clear that $(\alpha_i)_0^\infty$ is obtained by best extensions of the given segment.

THEOREM 6. *The set (α_{n+1}) of all numbers that provide best extensions of the absolutely monotonic segment $(\alpha_i)_0^n$ by one term fill a finite (closed) interval, the interval of best extension. In particular, the interval can reduce to a point.*

Suppose that both $\alpha_{n+1}^{(1)}$ and $\alpha_{n+1}^{(2)}$ provide best extensions of $(\alpha_i)_0^n$, i.e. that both segments

$$\alpha_0, \cdots, \alpha_n, \alpha_{n+1}^{(1)}; \quad \alpha_0, \cdots, \alpha_n, \alpha_{n+1}^{(2)}$$

are absolutely monotonic. Since multiplying such segments by $A > 0$, or adding them term by term, produces an absolutely monotonic segment, we have

$$\alpha_0, \alpha_1, \cdots, \alpha_n, \alpha_{n+1}, \quad \text{where} \quad \alpha_{n+1} = \frac{\gamma\alpha_{n+1}^{(1)} + \delta\alpha_{n+1}^{(2)}}{\gamma + \delta},$$

absolutely monotonic, and for arbitrary $\gamma \geqq 0$ and $\delta \geqq 0$ (not simultaneously zero) α_{n+1} runs through all numbers between $\alpha_{n+1}^{(1)}$ and $\alpha_{n+1}^{(2)}$.

On the other hand, the set $\{\alpha_{n+1}\}$ under discussion is clearly bounded, and consequently fills a finite interval; it is also clearly closed. We denote its endpoints by $\alpha'_{n+1} \leqq \alpha''_{n+1}$.

THEOREM 7. *If an absolutely monotonic segment $(\alpha_i)_0^n$ has an extremal polynomial $P_m(x)$, where $2 \leqq m \leqq n$, the best extensions of this segment are unique, beginning not later than the $(m+1)$th term.*

We make arbitrary best extensions of the given sequence by adding one term, two terms, etc.: this is theoretically always possible, according to the definition; we obtain an absolutely monotonic sequence $(\alpha_i)_0^\infty$ with the same extremal $P_m(x)$. According to Theorem 1, the integrator function $g(t)$ is a monotonic step function; its points $(\sigma_i)_1^s$ of discontinuity are determined by the condition $P_m(\sigma_i) = +1$, and the jumps $\delta_i > 0$ by

$$\sum_{i=1}^{s} \sigma_i^k \delta_i = \alpha_k \quad (k = 0, 1, \cdots, s-1),$$

where $s \leqq \frac{1}{2}m + 1 \leqq m$ (for $m \geqq 2$, p. 16). Thereafter all α_p $(p > m)$ are uniquely determined: $\alpha_p = \sum_{i=1}^{s} \sigma_i^p \delta_i$.

Now consider the case $m = 1$.

Suppose that in the segment α_0, α_1 we have $\alpha_0 > \alpha_1 > 0$. If this segment has an extremal polynomial, then it is either $1 - Cx$ or $1 - C(1-x)$; but it is easily verified that the segment-functional does not attain its norm for either of these polynomials. Therefore $P_m(x) = P_0(x) \equiv 1$. Finally, segments of the form α_0, α_0 or α_0, 0 (where $\alpha_0 > 0$) have respective extremal polynomials $1 - C(1-x)$ and $1 - Cx$, and each can be extended in a unique way.

Thus Theorem 7 is valid also for $m = 1$.

From now on we shall, for the sake of brevity, use the term *principal polynomial* of a segment for an extremal polynomial of smallest positive degree.

We shall call an absolutely monotonic segment *amorphous* if it has no extremal polynomial (except $P(x) \equiv 1$).

COROLLARY. *If $P_{\underline{n}}(x)$ is a principal polynomial for $(\alpha_i)_0^n$ then, by the corollary of Theorem* 1, *$P_n(x)$ has the structure of one of the types* I—IV (*p*. 15).

THEOREM 8. *An absolutely monotonic segment $(\alpha_i)_0^{n-1}$ with interval $[\alpha_n' < \alpha_n'']$ of best extension for α_n has an extremal polynomial of degree precisely n (not lower), if and only if either $\alpha_n = \alpha_n'$ or $\alpha_n = \alpha_n''$ (endpoint extension).*

In fact, suppose $\alpha_n' < \alpha_n < \alpha_n''$. Then the segment $(\alpha_i)_0^{n-1}$ is amorphous, since otherwise, by Theorem 7, its best extension would be unique; but if $P_n(\bar{\alpha}) = \alpha_0$ then $p_n = 0$.

Suppose $\alpha_n = \alpha_n''$; let the segment $(\alpha_i)_0^n$ define the functional A and suppose that the segment is amorphous. Replacing α_n'' by $\alpha_n'' + h$ $(h > 0)$ we obtain a non-absolutely-monotonic segment, defining a functional B with norm $N > \alpha_0$ and extremal (principal) polynomial $Q_n(x,h) = \sum_{i=0}^n q_i(h) x^i$.

Now $N(h) = B[Q_n(x,h)]$ tends to α_0 as $h \to 0$ (by Theorem 5); since the $q_i(h)$ are bounded, we can choose $h_p \to 0$ so that the limits $\lim_{h_p \to 0+} q_i(h_p)$ exist $(i = 0, 1, \cdots, n)$; then $\lim_{h_p \to 0+} Q_{\underline{n}}(x, h_p)$ also exists.

By hypothesis, this limit is 1; we shall show that this is impossible. In fact, for an extremal polynomial of a non-absolutely-monotonic segment there are at least two points $\xi_1(h)$ and $\xi_2(h)$ on $[0,1]$ such that $Q_n(\xi_1, h) = +1$ and $Q_n(\xi_2, h) = -1$. Suppose for definiteness that $1 \geqq \xi_2 > \xi_1 \geqq 0$. If $|q_i(h_p)| < \epsilon/n$ (for $i = 1, 2, \cdots, n$), then

$$Q_n(\xi_2, h_p) = q_0 + (q_0 \xi_2 + \cdots + q_n \xi_2^n) = \mathrm{I} + \mathrm{II};$$

but $\mathrm{I} \to +1$, $\mathrm{II} \to 0$, and $Q_n(\xi_2, h) \to -1$. The proof for $\alpha_n = \alpha_n'$ is similar. Thus in both cases there is an extremal polynomial of precise degree n.

COROLLARY 1. *If $\alpha_n = \alpha_n'$ or $\alpha_n = \alpha_n''$ further (best) extensions of the segment are unique. If $Q_n(x) = \sum_{k=0}^n q_k x^k$ is a principal polynomial of the segment, when $\alpha_n = \alpha_n''$ we have $q_n > 0$, and when $\alpha_n = \alpha_n'$ we have $q_n < 0$.*

This assertion is based on the following considerations: Let $\alpha_n = \alpha_n''$ and let the principal polynomial be $Q_n(x)$; if we replace α_n by $\alpha_n'' - \epsilon$ ($\epsilon > 0$; $\alpha_n'' - \epsilon > \alpha_n'$), let the resulting functional be $\bar{A}$. Then $\bar{A}(Q_n) = \alpha_0 - \epsilon q_n < \alpha_0$, since $\bar{A}$ does not have an extremal polynomial. Consequently $q_n > 0$.

COROLLARY 2. *If $\alpha_n' = \alpha_n'' = \alpha_n^*$ the segment $(\alpha_i)_0^{n-1}$ has an extremal polynomial.*

Suppose the contrary. Replacing α_n^* by $\alpha_n^* + h$ and then by $\alpha_n^* - h$, we obtain two non-absolutely-monotonic functionals B_1 and B_2 with extremal polynomials $Q_n^{(1)}(x, h)$ and $Q_n^{(2)}(x, h)$. By considerations similar to those above, we can show that (as $h \to 0$) the segment $\alpha_0, \alpha_1, \cdots, \alpha_{n-1}, \alpha_n^*$ has two (principal!) extremal polynomials $P_n^{(1)}(x)$ and $P_n^{(2)}(x)$, with $p_n^{(1)} > 0$ and $p_n^{(2)} < 0$, which contradicts the corollary of Theorem 1, according to which the leading coefficient has a fixed sign. Thus $p_n^{(1)} = p_n^{(2)} = 0$.

We now give some examples of absolutely monotonic segments.

EXAMPLE 11. The segment consisting of the single number $\alpha_0 > 0$ has the interval $[\alpha_1', \alpha_1'']$ of best extension, where $\alpha_1' = 0$ and $\alpha_1'' = \alpha_0$; the corresponding principal polynomials are $1 - Cx$ and $1 - C(1 - x)$.

EXAMPLE 12. The segment α_0, α_1 with $\alpha_0 > \alpha_1 > 0$ has as its interval of best extension the interval with endpoints $\alpha_2' = \alpha_1^2/\alpha_0$ and $\alpha_2'' = \alpha_1$; the corresponding extremal polynomials are

$$P_2(x) = 1 - C\left(x - \frac{\alpha_1}{\alpha_0}\right)^2 \quad \text{and} \quad P_2(x) = 1 - Cx(1 - x).$$

EXAMPLE 13. The segment $1, q, q^2, \cdots, q^n = (\alpha_i)_0^n$ with $0 < q < 1$ and $n \geqq 2$ has the unique best extension $\alpha_{n+1} = q^{n+1}$; the principal polynomials of the segment are $1 - C(x - q)^2$, and $q^2 = \alpha_2'$; $q = \alpha_2''$; $q^3 = \alpha_3' = \alpha_3''$, etc.

Going over to the problem of the actual (best) extension of a given absolutely monotonic segment, we make some introductory remarks.

If $(\alpha_i)_0^\infty$ is an absolutely monotonic segment and its integrator function $g(t)$ is a step function with a finite number of jumps, then $(\alpha_i)_1^\infty$ and

$(\alpha_{i,1})_0^\infty$, where $\alpha_{i,1} = \alpha_i - \alpha_{i+1}$, are also absolutely monotonic and the corresponding integrator functions $g_1(t)$ and $g_2(t)$ are also step functions with a finite number of jumps. If $(\sigma_i)_1^s$ are the abscissas of the jumps of $g(t)$, with jumps of amounts $(\delta_i)_1^s$, then $g_1(t)$ has the same discontinuities, except for $t = 0$ if $g(t)$ has a jump at that point, and the amount of the jump of $g_1(t)$ is $(\sigma_i \cdot \delta_i)$. In fact,

$$\alpha_1 = \int_0^1 dg_1(t) = \sum_{i=1}^s \sigma_i\delta_i, \cdots, \alpha_n = \int_0^1 t^{n-1} dg_1(t) = \sum_{i=1}^s \sigma_i^{n-1}(\sigma_i\delta_i), \cdots.$$

Consequently the degree of a principal polynomial of $(\alpha_i)_1^\infty$ can be one less than the degree of a principal polynomial of $(\alpha_i)_0^\infty$, provided that $\sigma_1 = 0$.

Similarly $g_2(t)$ has discontinuities (σ_i) with jumps $(1 - \sigma_i)\,\delta_i$, since

$$\alpha_{k,1} = \int_0^1 t^k(1-t)\,dg(t) = \sum_{i=1}^s \sigma_i^k(1-\sigma_i)\,\delta_i.$$

Consequently if $g(t)$ has a jump at $t = 1$, the degree of an extremal polynomial for $(\alpha_{k,i})_0^\infty$ is reduced by 1.

Let the segment $(\alpha_i)_0^n$ have $P_{\underline{n}}(x)$ as principal polynomial. Let $(\sigma_i)_1^s$ be the nodes of $P_{\underline{n}}(x)$, and let $R_s(x) = \prod_1^s (x - \sigma_i)$ be its resolvent (p. 16).

Extending the segment (in the best way), uniquely, we obtain $(\alpha_i)_0^\infty$. If A is the functional determined by this sequence, we evidently have the following equation, where $Q(x)$ is an arbitrary polynomial: $A(R \cdot Q) = 0$, since we have $A(P) = \sum_{i=1}^s \delta_i P(\sigma_i)$.

THEOREM 9. *Let* $(\alpha_i)_0^{n-1}$, *with* $n \geqq 3$, *have no extremal polynomial; let* $(\sigma_i)_1^s$ *be the nodes of a principal polynomial of the segment* $(\alpha_i)_0^n$, *where* α_n *is either* α_n' *or* α_n''; *then*:

1. *If* $2s = n$, *then* α_n' *is determined by*

$$\delta(\alpha_0, \alpha_n') = 0, \quad \textit{where} \quad \delta(\alpha_0, \alpha_n') = \begin{vmatrix} \alpha_0 & \cdots & \alpha_s \\ \alpha_1 & \cdots & \alpha_{s+1} \\ \cdot & \cdot & \cdot \\ \alpha_s & \cdots & \alpha_n' \end{vmatrix}; \quad (\alpha_n' = \alpha_{2s})$$

2. *If* $2s - 1 = n$ *then* α_n' *is determined by*

$$\delta(\alpha_1, \alpha_n') = 0, \quad \textit{where} \quad \delta(\alpha_1, \alpha_n') = \begin{vmatrix} \alpha_1 & \cdots & \alpha_s \\ \alpha_2 & \cdots & \alpha_{s+1} \\ \cdot & \cdot & \cdot \\ \alpha_s & \cdots & \alpha_n' \end{vmatrix}; \quad (\alpha_n' = \alpha_{2s-1}).$$

3. *If* $2s-1=n$, *then* $\alpha''_n=\alpha_{n-1}-\alpha'_{n-1,1}$, *where* $\alpha'_{n-1,1}=\beta'_{n-1}$ $(\alpha_{k,1}=\beta_k)$ *is determined by*

$$\delta(\beta_0,\beta'_{n-1})=0,\ \textit{where}\ \delta(\beta_0,\beta'_{n-1})=\begin{vmatrix}\beta_0\cdots\beta_{s-1}\\ \beta_1\cdots\beta_s\\ \cdot\ \cdot\ \cdot\ \cdot\\ \beta_{s-1}\cdots\beta'_{n-1}\end{vmatrix};\ (\beta'_{n-1}=\beta'_{2s-2}=\alpha'_{n-1,1}).$$

4. *If* $2s-2=n$, *then* $\alpha''_n=\alpha_{n-1}-\alpha'_{n-1,1}$ (*here* $\alpha'_{n-1,1}=\beta'_{n-1}$), *and is determined by*

$$\delta(\beta_1,\beta'_{n-1})=0,\quad \textit{where}\quad \delta(\beta_1,\beta'_{n-1})=\begin{vmatrix}\beta_1\cdots\beta_{s-1}\\ \beta_2\cdots\beta_s\\ \cdot\ \cdot\ \cdot\ \cdot\\ \beta_{s-1}\cdots\beta'_{n-1}\end{vmatrix}.$$

In each of these four cases we denote an extremal polynomial of the segment $(\alpha_i)_0^n$ by $Q_{\underline{n}}(x)=\sum_0^n q_ix^i$. By the remark on p. 16, no other cases are possible.

1. If $\alpha_n=\alpha'_n$ then $q_n<0$ (Corollary 1 of Theorem 8), the segment $(\alpha_i)_0^n$ has, when n is even, only interior nodes $(\sigma_i)_1^s$, and $2s=n$. The coefficients of the resolvent

$$R_s(x)=\sum_0^s r_ix^i=\prod_1^s(x-\sigma_i)$$

satisfy the system of equations (remark on p. 25)

$$\begin{aligned}A[R_s(x)]&=\alpha_0r_0+\cdots+\alpha_{s-1}r_{s-1}+\alpha_sr_s=0\\ A[xR_s(x)]&=\alpha_1r_0+\cdots+\alpha_sr_{s-1}+\alpha_{s+1}r_s=0\\ &\cdots\cdots\cdots\cdots\cdots\\ A[x^sR_s(x)]&=\alpha_sr_0+\cdots+\alpha_{2s-1}r_{s-1}+\alpha'_nr_s=0.\end{aligned}\tag{20}$$

Since $r_s=1$, the determinant $\delta(\alpha_0,\alpha'_n)$ is automatically 0; if we show that the minor of α'_n is zero, i.e. $\delta(\alpha_0,\alpha_{n-2})=0$, then by the preceding reasoning we conclude that $\delta(\alpha_0,\alpha'_{n-2})=0$ also, where $\alpha'_{n-2}\neq\alpha_{n-2}$, since the segment $\alpha_0,\alpha_1,\cdots,\alpha_{n-2}$ has no extremal polynomial, and so on; finally we necessarily have $\alpha_0=0$, which is contradictory.

Thus α'_n is uniquely determined by $\delta(\alpha_0,\alpha'_n)=0$.

2. For odd n and $\alpha_n=\alpha'_n$ we have $q_n<0$; $\sigma_1=0$, $\sigma_s<1$, and $2s-1=n$. Consider the segment $\alpha_1,\alpha_2,\cdots,\alpha_{n-1},\alpha'_n$; by a previous remark (pp. 24-25) this segment has an extremal polynomial with the same nodes except

for $\sigma_1 = 0$, i.e. all its nodes are interior and α'_n is determined, as in the first case, by $\delta(\alpha_1, \alpha'_n) = 0$.

3. If $\alpha_n = \alpha''_n$ then $q_n > 0$; when n is odd we must have $2s - 1 = n$; $\sigma_s = 1$; $\sigma_1 > 0$.

Consider the segment $\alpha_{0,1}, \alpha_{1,1}, \cdots, \alpha_{n-2,1}, \alpha'_{n-1,1}$ $(\alpha'_{n-1,1} = \alpha_{n-1} - \alpha''_n)$; the nodes of this segment are the same, except for $\sigma_s = 1$; hence all the nodes are interior, and with $\alpha'_{n-1,1} = \beta'_{n-1}$, we obtain (as in case 1) β'_{n-1} from $\delta(\beta_0, \beta'_{n-1}) = 0$, and then $\alpha''_n = \alpha_{n-1} - \alpha'_{n-1,1}$.

4. Finally, when $\alpha_n = \alpha''_n$ and n is even we must have

$$q_n > 0; \quad 2s - 2 = n; \quad \sigma_1 = 0; \quad \sigma_s = 1.$$

The segment $\alpha_{1,1}, \alpha_{2,1}, \cdots, \alpha_{n-2,1}, \alpha_{n-1,1}$ lacks these two nodes but has all the rest; its extremal polynomial is of degree $n - 2$; since $\alpha_{n-1,1} = \alpha_{n-1} - \alpha''_n$, in this case we also have $\alpha_{n-1,1} = \alpha'_{n-1,1}$. Putting $\alpha'_{n-1,1} = \beta'_{n-1}$, we find it from $\delta(\beta_1, \beta'_{n-1}) = 0$, and then $\alpha''_n = \alpha_{n-1} - \alpha'_{n-1,1}$. This completes the proof of Theorem 9.

Thus for any amorphous segment $(\alpha_i)_0^{n-1}$ each of the numbers α'_n and $\alpha''_n > \alpha'_n$ makes a determinant of the form δ vanish. Conversely, if one of these determinants is zero, α_n is at the corresponding endpoint of the interval of best extension. If $\alpha'_n = \alpha''_n$ the principal polynomial is of degree less than n.

COROLLARY. *If an absolutely monotonic segment $(\alpha_i)_0^n$ is given, the degree m of a principal polynomial and the coefficients $(r_i)_0^{n-1}$ of its resolvent are easily obtained in the following way: arrange the determinants δ according to increasing order (as we shall see below, none of them can be negative); if none of them is zero, the segment has no extremal polynomial. Let the vanishing determinant of lowest order be $\delta(\alpha_0, \alpha_{2k})$; then $\alpha_{2k} = \alpha'_{2k}$, $m = 2k$, and all k nodes are interior; if the first determinant that is zero is $\delta(\alpha_1, \alpha_{2k-1})$ then $\alpha_{2k-1} = \alpha'_{2k-1}$ and $m = 2k - 1$; there are k nodes, one of which is 0. If the first determinant that is zero is $\delta(\alpha_{0,1}, \alpha_{2k-2,1})$ then $\alpha_{2k-1} = \alpha''_{2k-1}$, $m = 2k - 1$; one of the k nodes is 1. Finally, if the first determinant that is zero is $\delta(\alpha_{1,1}, \alpha_{2k-3,1})$ then $m = 2k - 2$, $\alpha_{2k-2} = \alpha''_{2k-2}$; both 0 and 1 occur among the k nodes.*

In all four cases the coefficients r_i are obtained uniquely from the elements of the segment $(\alpha_i)_0^m$ by a system of equations of the form (20).

We now formulate a criterion for a segment to be absolutely monotonic; we exclude the trivial case $\alpha_0, \alpha_1, \cdots, \alpha_1$, where $\alpha_0 \geqq \alpha_1 \geqq 0$ and the difference table contains zeros.

Theorem 10 (First criterion). *A necessary condition for the segment $(\alpha_i)_0^n$ to be absolutely monotonic is that the following conditions (using the notation of Theorem 9) are satisfied:*

1) $\alpha_0 > \alpha_1 > \alpha_2 > 0$;
2) $\delta(\alpha_0, \alpha_{2k}) \geqq 0, k = 1, 2, \cdots, n/2$ *or* $(n-1)/2$;
   $\delta(\alpha_1, \alpha_{2k+1}) \geqq 0, k = 1, 2, \cdots, \frac{1}{2}n - 1$ *or* $\frac{1}{2}(n-1)$;
3) $\delta(\beta_0, \beta_{2k}) \geqq 0, k = 1, 2, \cdots, \frac{1}{2}(n-1)$ *or* $\frac{1}{2}n - 1$;
   $\delta(\beta_1, \beta_{2k+1}) \geqq 0, k = 1, 2, \cdots, \frac{1}{2}(n-1) - 1$ *or* $\frac{1}{2}n - 1$

(*these are all the determinants of type δ that can be formed*).

If all these determinants are positive, the condition is also sufficient.

Necessity. Let $(\alpha_i)_0^n$ be an absolutely monotonic segment. In the determinants of type $\delta(\alpha_0, \alpha_{2k})$, put $\alpha_{2k} = \alpha'_{2k} + h_{2k}$, where we must have $h_{2k} \geqq 0$, since either α'_{2k} is the left-hand endpoint of an interval of best extension (then $h_{2k} \geqq 0$) or α_{2k} is a unique best extension (then $h_{2k} = 0$). We obtain the recurrent formula $\delta(\alpha_0, \alpha_{2k}) = h_{2k} \cdot \delta(\alpha_0, \alpha_{2k-2})$; but since $\delta(\alpha_0, \alpha_2) = h_2\alpha_0 \geqq 0$, it follows that $\delta(\alpha_0, \alpha_{2k}) \geqq 0$. In the same way $\delta(\alpha_1, \alpha_{2k+1}) \geqq 0$, since $(\alpha_i)_1^n$ is also an absolutely monotonic segment.

Similar considerations hold for determinants of the types $\delta(\beta_0, \beta_m)$ and $\delta(\beta_1, \beta_m)$, since the segments $(\beta_i)_0^{n-1}$ and $(\beta_i)_1^{n-1}$ are also absolutely monotonic.

Sufficiency. Let all the determinants in question be positive for $(\alpha_i)_0^n$. Supposing that $(\alpha_i)_0^{k-1}$ is absolutely monotonic, we show that $(\alpha_i)_0^k$ is also absolutely monotonic. By Theorem 9, $(\alpha_i)_0^{k-1}$ is an amorphous segment, i.e. $\alpha'_k < \alpha''_k$.

1) Suppose that $a_k = \alpha'_k - h < \alpha'_k$. If k is even, $\delta(\alpha_0, \alpha_k) = -h\delta(\alpha_0, \alpha_{k-2}) < 0$, which is impossible. If k is odd, $\delta(\alpha_1, \alpha_k) = -h\delta(\alpha_1, \alpha_{k-3}) < 0$, which is also impossible.

2) Suppose that $\alpha_k = \alpha''_k + h > \alpha''_k$; we form the first differences $\beta_0, \beta_1, \cdots, \beta_{k-1}$; then $\beta_{k-1} = \beta'_{k-1} - h$, and case 2 is reduced to case 1.

Thus $(\alpha_i)_0^k$ is absolutely monotonic if $(\alpha_i)_0^{k-1}$ is. But the segment α_0, α_1 is absolutely monotonic and has an interval of best extension. In fact, $\alpha'_2 = \alpha_1^2/\alpha_0$ since $(\alpha_0, \alpha'_2) = 0$; $\alpha''_2 = \alpha_1$; in addition, $\delta(\alpha_0, \alpha_2) > 0$ and $\alpha_{1,1} = \alpha_1 - \alpha_2 > 0$ by the hypotheses of the theorem.

Thus $(\alpha_i)_0^n$ is an absolutely monotonic segment, and the first criterion is established. Note that if one of the determinants δ is zero, the criterion is not sufficient.

Theorem 11 (Second criterion). *Let the segment $(\alpha_i)_0^n$ be given and let m be the smallest integer, $3 \leqq m \leqq n$, such that among the four deter-*

minants $\delta(\alpha_0, \alpha_m)$, $\delta(\alpha_1, \alpha_m)$, $\delta(\beta_0, \beta_{m-1})$, $\delta(\beta_1, \beta_{m-1})$ *one of the two that can be zero, is zero; i.e., the determinants of this form are positive if m is replaced by* $k < m$; *then a necessary and sufficient condition that the segment* $(\alpha_i)_0^n$ *is absolutely monotonic is that*

$$\begin{aligned}
&\text{if } \delta(\alpha_0, \alpha_m) = 0, \text{ then all determinants of the form } \delta(\alpha_k, \alpha_{k+m}) \text{ are zero;}\\
&\text{if } \delta(\alpha_1, \alpha_m) = 0, \text{ then all determinants of the form } \delta(\alpha_{k+1}, \alpha_{k+m}) \text{ are zero;}\\
&\text{if } \delta(\beta_0, \beta_{m-1}) = 0, \text{ then all determinants of the form } \delta(\beta_k, \beta_{k+m-1}) \text{ are zero;}\\
&\text{if } \delta(\beta_1, \beta_{m-1}) = 0, \text{ then all determinants of the form } \delta(\beta_{k+1}, \beta_{k+m-1}) \text{ are zero;}\\
&\qquad\qquad(\text{in all cases } k > 0).
\end{aligned}$$

Necessity. Let $(\alpha_i)_0^n$ be an absolutely monotonic segment and $\delta(\alpha_0, \alpha_m) = 0$. Consider the segments $(\alpha_i)_0^m, (\alpha_i)_1^{m+1}, \cdots, (\alpha_i)_{n-m}^n$; they are all absolutely monotonic; by Theorem 9, $\alpha_m = \alpha_m'$, and all these segments have the same extremal (principal) polynomial $P_{\underline{m}}(x)$. Consequently $\delta(\alpha_k, \alpha_{k+m}) = 0$ also. The proofs for the other three cases are similar.

Sufficiency. Suppose for definiteness that the number m is even. Given $\delta(\alpha_0, \alpha_m) = 0$, then by the hypothesis of the theorem ($m = \min$) the segment $(\alpha_i)_0^{m-1}$ is absolutely monotonic and has no extremal polynomial (Theorem 10).

In addition, $\alpha_m = \alpha_m'$ and the segment $(\alpha_i)_1^m$ is absolutely monotonic, with principal polynomial $Q_{\underline{m}}(x)$; consequently the segment can be extended (in the best way) uniquely, for any number of terms.

Let the segment so extended be $\alpha_0, \alpha_1, \cdots, \alpha_m, \alpha_{m+1}^*, \cdots, \alpha_n^* = (\gamma_i)_0^n$; a principal polynomial $Q_{\underline{m}}(x)$ for it is also a principal polynomial for the segment $\gamma_1, \gamma_2, \cdots, \gamma_{m+1}$, and then, by Theorem 9, γ_{m+1} is uniquely determined by $\delta(\gamma_1, \gamma_{m+1}) = 0$, and consequently $\gamma_{m+1} = \alpha_{m+1}$; in the same way we show that $\gamma_{m+2} = \alpha_{m+2}, \cdots, \gamma_n = \alpha_n$.

Thus $(\alpha_i)_0^n$ is an absolutely monotonic segment. (For the other cases the proof is carried out by using differences.)

Remark. If $m = 2$, and by hypothesis $\delta(\alpha_0, \alpha_2) = 0$, the segment $\alpha_0, \alpha_1, \alpha_2$ is a geometric progression (p. 24), with $\rho = \alpha_1/\alpha_0$, and an extremal polynomial $Q_2(x)$ has the form $Q_2(x) = 1 - C(x - \alpha_1/\alpha_0)^2$.

Example 14. $(\alpha_i)_0^n = 1, q, q^2, \cdots, q^n$ $(0 < q < 1)$ is an absolutely monotonic segment (Example 13).

We check the criteria for this segment. We have $\delta(\alpha_0, \alpha_2) = 0$, $\delta(\alpha_0, \alpha_4) = 0$, etc.; likewise $\delta(\alpha_1, \alpha_3) = 0$, $\delta(\alpha_1, \alpha_5) = 0$, etc. It follows that $q^2 = \alpha_2'$ and the extension is unique starting with α_3. If we put $\alpha_{n+1} = x$, then $\delta(\alpha_0, x) = 0$ or $\delta(\alpha_1, x) = 0$ for each x, and the first criterion does not provide an extension of the segment.

The second criterion requires $\delta(\alpha_{m-2}, \alpha_m) = 0$ and provides the desired extension by the equation $\delta(q^{n-1}, x) = 0$, whence $x = q^{n+1}$.

EXAMPLE 15. $(\alpha_i)_0^{n-1} = 1, 1/2, 1/3, \cdots, 1/n$. This is an absolutely monotonic segment (Example 8), and is amorphous.

1. Let n be even; we find the interval (α_n', α_n'') of best extension. First we determine α_n'. The resolvent of the extremal polynomial $R_{n/2}(x)$ of the segment to be extended can, by (20), be represented in the form

$$C_0 \cdot R_{n/2}(x) = \begin{vmatrix} 1 \frac{1}{2} \cdots \frac{1}{n/2+1} \\ \frac{1}{2}\, \frac{1}{3} \cdots \frac{1}{n/2+2} \\ \cdot\ \cdot\ \cdot\ \cdot\ \cdot\ \cdot\ \cdot\ \cdot \\ \frac{1}{n/2} \quad \cdots \quad \frac{1}{n} \\ 1 \quad x \cdots x^{n/2} \end{vmatrix}, \quad \text{where} \quad C_0 = \delta\left(1, \frac{1}{n-1}\right).$$

Then if A is the functional determined by $(\alpha_i)_0^{n-1}$, we have

$$C_0 \cdot A(R_{n/2}) = C_0 \int_0^1 R_{n/2}(x)\,dx = 0;$$

$$C_0 A(xR_{n/2}) = C_0 \int_0^1 xR_{n/2}(x)\,dx = 0, \cdots$$

$$C_0 A(x^{n/2-1}R_{n/2}) = C_0 \int_0^1 x^{n/2-1}R_{n/2}(x)\,dx = 0.$$

Thus the polynomial $C_0R_{n/2}(x)$ is orthogonal on $[0,1]$ to every polynomial of degree not exceeding $\frac{1}{2}n - 1$ (the functional is not defined beyond this).

Consequently $R_{n/2}(x)$ is the Legendre polynomial [2]; it has the form

$$R_{n/2}(x) = (-1)^{n/2}\frac{(n/2)!}{n!}\,\frac{d^{n/2}[x^{n/2}(1-x)^{n/2}]}{dx^{n/2}}.$$

To calculate α_n' we put $\alpha_n' = \alpha_n - h = 1/(n+1) - h$ in the determinant $\delta(\alpha_0, \alpha_n')$ (criterion for α_n'). Then

$$C_0 = \delta(\alpha_0, \alpha_{n-2});$$

$$\delta(\alpha_0, \alpha_n') = \delta(\alpha_0, \alpha_n) - h\delta(\alpha_0, \alpha_{n-2}) = 0;$$

$$\delta(\alpha_0, \alpha_2) = 1/12;$$

$$h = \frac{\delta(\alpha_0, \alpha_n)}{\delta(\alpha_0, \alpha_{n-2})}; \quad \delta(\alpha_0, \alpha_n) = \int_0^1 x^{n/2} R_{n/2}(x)\, dx \cdot \delta(\alpha_0, \alpha_{n-2});$$

$$h = \int_0^1 x^{n/2} R_{n/2}(x)\, dx = (-1)^{n/2} \frac{(n/2)!}{n!} \int_0^1 x^{n/2} \frac{d^{\,n/2}[x^{n/2}(1-x)^{n/2}]}{dx^{n/2}}\, dx.$$

By integration by parts we obtain

$$h = \frac{(n/2!)^2}{n!} \int_0^1 x^{n/2}(1-x)^{n/2} dx$$

$$= \frac{(n/2!)^2}{n!} B\left(\frac{n}{2}+1, \frac{n}{2}+1\right) = \frac{(n/2!)^4}{(n+1)(n!)^2}.$$

[$B(p,q)$ is the Euler integral.] Thus

$$\alpha_n' = \frac{1}{n+1}\left[1 - \frac{(n/2!)^4}{(n!)^2}\right].$$

2. Now we find α_n' for odd n. Here $R_{(n+1)/2}(x)$ can be written in the form

$$C_0 \cdot R_{(n+1)/2}(x) = \begin{vmatrix} \frac{1}{2} & \frac{1}{3} & \cdots & \frac{1}{(n+1)/2+1} \\ \cdot & \cdot & \cdot & \cdot \\ \frac{1}{(n+1)/2} & & \cdots & \frac{1}{n} \\ x & x^2 & \cdots & x^{(n+1)/2} \end{vmatrix}, \quad \text{where} \quad C_0 = \delta(\alpha_1, \alpha_{n-2}).$$

Then

$$\int_0^1 R_{(n+1)/2}(x)\, dx = 0;$$

$$\int_0^1 x R_{(n+1)/2}(x)\, dx = 0; \cdots; \int_0^1 x^{(n-1)/2-1} R_{(n+1)/2}(x)\, dx = 0.$$

Thus the resolvent is of the form $xR_{(n-1)/2}(x)$ and is orthogonal on $[0,1]$ to all polynomials of degree not exceeding $\frac{1}{2}(n-1) - 1$.

Put

$$R_{(n+1)/2}(x) = k \cdot \frac{d^{(n-1)/2} P_n(x)}{dx^{(n-1)/2}},$$

then $x = 1$ is a root of $P_n(x)$ of multiplicity $\frac{1}{2}(n-1)$, and $x = 0$ is a root of multiplicity $\frac{1}{2}(n+1)$.

Consequently

$$P_n(x) = x^{(n+1)/2}(1-x)^{(n-1)/2};$$

$$R_{(n+1)/2}(x) = (-1)^{(n-1)/2}\frac{((n+1)/2)!}{n!}\frac{d^{(n-1)/2}[x^{(n+1)/2}(1-x)^{(n-1)/2}]}{dx^{(n-1)/2}}.$$

In addition,

$$\delta(\alpha_1, \alpha_n') = \delta(\alpha_1, \alpha_n) - h\delta(\alpha_1, \alpha_{n-2}) = 0;$$

$$\delta(\alpha_1, \alpha_n) = \delta(\alpha_1, \alpha_{n-2})\int_0^1 x^{(n-1)/2}R_{(n+1)/2}(x)\,dx;$$

$$\delta(\alpha_1, \alpha_5) = \delta(\alpha_1, \alpha_3)\int_0^1 x^2 R_3(x)\,dx; \quad \delta(\alpha_1, \alpha_3) = \frac{1}{72};$$

$$h = \frac{\delta(\alpha_1, \alpha_n)}{\delta(\alpha_1, \alpha_{n-2})} = \int_0^1 x^{(n-1)/2}R_{(n+1)/2}(x)\,dx$$

$$= (-1)^{(n-1)/2}\frac{((n+1)/2)!}{n!}\int_0^1 x^{(n-1)/2}\frac{d^{(n-1)/2}[x^{(n+1)/2}(1-x)^{(n-1)/2}]}{dx^{(n-1)/2}}dx$$

$$= \frac{((n-1)/2)!\,((n+1)/2)!}{n!}B\left(\frac{n+3}{2}, \frac{n+1}{2}\right).$$

3. Similarly we find α_n'' for even n. The resolvent

$$R_{n/2+1}(x) = (-1)^{n/2}\frac{(n/2+1)!}{n!}\frac{d^{n/2-1}[x^{n/2}(1-x)^{n/2}]}{dx^{(n-1)/2}};$$

$$\alpha_n'' = \frac{1}{n+1} + h;$$

$$h = \frac{(n/2!)^2(n/2+1)!\,(n/2-1)!}{(n+1)!\cdot n!}.$$

4. In the same way we find α_n'' for odd n.

$$R_{(n+1)/2}(x) = (-1)^{(n+1)/2}\frac{((n+1)/2)!}{n!}\frac{d^{(n-1)/2}[x^{(n-1)/2}(1-x)^{(n+1)/2}]}{dx^{(n-1)/2}};$$

$$\alpha_n'' = \frac{1}{n+1} + h;$$

$$h = \frac{((n-1)/2!)^2((n+1)/2!)^2}{n!\,(n+1)!}.$$

The method of extending an absolutely monotonic segment shows that in general the ends of the interval of best extension are rational functions of the parameters of the segment.

REMARK. The theory of best extension of an absolutely monotonic segment can be extended to any term α_k $(k < n)$; thus, a segment $(\alpha_i)_0^n$ $(i \neq k)$ with one missing term is called absolutely monotonic if there is a value of α_k which makes it absolutely monotonic. One can prove theorems similar to Theorems 6—11, but we shall omit the proofs.

§3. The non-absolutely-monotonic segment and its extremal polynomials

Suppose that the segment $(\mu_k)_0^n$ is not absolutely monotonic and let N_n be the norm of the corresponding functional F_n. Let $Q_p(x)$ be one of its extremal polynomials (such polynomials always exist, and $0 < p \leqq n$), i.e. $Q_p(\bar{\mu}) = +N_n$.

THEOREM 12 [8]. *Each non-absolutely-monotonic segment $(\mu_i)_0^n$ has a unique best extension, i.e. there is just one number μ_{n+1}^* such that the segment $\mu_0, \mu_1, \cdots, \mu_n, \mu_{n+1}^*$ defines a functional F_{n+1} with the same norm N_n.*

Let $Q_n(x)$ be an arbitrary extremal polynomial of the segment $(\mu_i)_0^n$ and form an arbitrary sequence of best extensions of the given segment by one term, by two, etc. (this is possible by the Hahn-Banach Theorem):

$$\mu_0, \mu_1, \cdots, \mu_n, \mu_{n+1}, \cdots, \mu_p, \cdots. \tag{21}$$

We obtain a moment sequence, i.e.

$$\mu_p = \int_0^1 t^p dH(t); \quad p = 0, 1, 2, \cdots,$$

for which $Q_n(x)$ is an extremal polynomial. Then by Theorem 2, $H(t)$ is a step function with a finite number of discontinuities, at the points $(\sigma_i)_1^s$; let the corresponding jumps of $H(t)$ be $(\delta_i)_1^s$. Then we must have $Q_n(\sigma_i) = \operatorname{Sgn} \delta_i$.

For any extension of the segment $(\mu_k)_0^n$ with preservation of its norm, the σ_i must be points of deviation of $Q_n(x)$ on $[0, 1]$, i.e. $|Q_n(\sigma_i)| = 1$. Let $(\sigma_i')_1^{s_1}$ be all the points of deviation of $Q_n(x)$ on $[0, 1]$. Clearly $2 \leqq s \leqq s_1 \leqq n + 1$.

Then the $(\delta_i)_1^s$ at the points σ_i' are determined uniquely by the system of s_1 equations $\sum_{i=1}^{s_1} \sigma_i'^k \delta_i = \mu_k$ $(k = 0, 1, \cdots, s_1 - 1)$ with a Vandermonde determinant. (If a point σ_i' is not a point of discontinuity of $H(t)$, its weight δ is zero.)

By the definition of (δ_i), the μ_p with $p \geqq s_1$ are uniquely defined by

$$\mu_p = \sum_{i=1}^{s} \sigma_i^p \delta_i.$$

Therefore the segment $(\mu_k)_0^n$ has a unique sequence of best extensions, which we denote by

$$\mu_0, \mu_1, \cdots, \mu_n, \overset{*}{\mu}_{n+1}, \overset{*}{\mu}_{n+2}, \cdots,$$

here each $\mu_{n+1}^*, \cdots$ is unique, and the extremal polynomials $\{Q_n(x)\}$ are all such that the points $(\sigma_i)_1^s$ of discontinuity of the unique $H(t)$ are some of the points of deviation of each polynomial (not necessarily all of them).

Corollary 1. *To each non-absolutely-montonic segment-functional* $(\mu_k)_0^n$ *there corresponds a definite set* $(\sigma_i)_1^s$, *the points of discontinuity of the step-function* $H(t)$, *where* $0 \leqq \sigma_i \leqq 1$.

If we attach to each point σ_i the sign of the corresponding jump of $H(t)$, we obtain a "distribution" $(\overset{\pm}{\sigma}_i)_1^s$, which we call the *faithful* distribution of the segment $(\mu_k)_0^n$. (Such a distribution necessarily contains some numbers to which different signs are attached.)

Thus each non-absolutely-monotonic segment $(\mu_k)_0^n$ has a unique faithful distribution $(\overset{\pm}{\sigma}_i)_1^s$, in terms of which its terms can be expressed as follows:

$$\mu_k = \sum_{i=1}^{s} \sigma_i^k \cdot \delta_i \quad (k = 0, 1, \cdots, n). \tag{22}$$

We call this decomposition in terms of the nodes (σ) a faithful decomposition.

Corollary 2. *If* $(\mu_k)_0^n$ *is a non-absolutely-monotonic segment, and* $(\overset{\pm}{\sigma}_i)_1^s$ *is its faithful distribution, then there is a reduced polynomial* $Q_n(x)$ *for which* $Q_m(\overset{+}{\sigma}) = 1$ *and* $Q_m(\overset{-}{\sigma}) = -1$; $Q_m(x)$ *is an extremal polynomial for the segment, and* $m \leqq n$.

Because of possible special cases, we call the faithful distribution $(\overset{\pm}{\sigma}_i)_1^s$ of a segment a polynomial distribution of degree m provided that m is the degree of a principal (i.e., of lowest degree) extremal polynomial of the segment; $1 \leqq m \leqq n$.

Theorem 13. *Let* $(\mu_k)_0^n$ *be any segment, and* $(C_i)_1^{n+1}$ *arbitrary points with* $0 \leqq C_1 < C_2 < \cdots < C_{n+1} \leqq 1$. *Then if we solve the system of* $n+1$ *equations in* $n+1$ *unknowns* x_i,

(23) $$\sum_{i=1}^{n+1} C_i^k x_i = \mu_k \quad (k = 0, 1, 2, \cdots, n),$$

the norm of the segment satisfies

(24) $$N_n \leqq \sum_{i=1}^{n+1} |\Delta_i|,$$

where Δ_i are the roots of the system.

In fact, after we solve (23) the set (C_i) becomes the distribution $(\overset{\pm}{C}_i)_1^{n+1}$ if we attach to each C_i the sign of Δ_i and discard the C_i for which $\Delta_i = 0$. Then we can obviously construct a reduced polynomial $Q(x)$ for which $Q(\overset{+}{C}) = +1$ and $Q(\overset{-}{C}) = -1$ (by the process indicated on pp. 17).

If the lowest possible degree for such a polynomial is $p \leqq n$, then $(\overset{\pm}{C}_i)$ is the faithful distribution of $(\mu_i)_0^n$ and $Q_p(\bar{\mu}) = \sum |\Delta_i| = N_n$.

On the other hand, if $p > n$ then $N_n < Q_p(\bar{\mu}) = \sum |\Delta_i|$, and the theorem is established.

COROLLARY 1. *If all $\Delta_i \geqq 0$, the segment $(\mu_i)_0^n$ is absolutely monotonic; and if the number of points for which $\Delta_i > 0$ is $m \leqq n/2$, the corresponding C_i are the faithful nodes of the segment (here we must distinguish the four forms discussed above).*

COROLLARY 2. *If $\{P_n(x)\}$ is a set of reduced polynomials, then*

$$|P_n(\bar{\mu})| \leqq N_n \leqq \sum_1^{n+1} |\Delta_i|.$$

Therefore the faithful representation $(\overset{\pm}{\sigma}_i)_1^n$ of a segment has the property that if we replace (C_i) by (σ_i) in (23), the sum $\sum_1^{n+1} |x_i|$ attains its minimum.

For a reduced $P_n(x)$ let $(\sigma_i)_1^s$ be its points of deviation on $[0, 1]$; then if $P_n(\sigma_i) = +1 \ (-1)$, we obtain $(\overset{\pm}{\sigma}_i)_1^s$, the distribution of the polynomial.

COROLLARY 3. *If the distribution $(\overset{\pm}{\sigma}_i)_1^s$ of a polynomial $P_n(x)$ is such that, for a given segment $(\mu_k)_0^n$, replacing (C_i) by (σ_i) in (23) leads to a consistent system ($s \leqq n+1$), with roots $x_i = \delta_i$ such that either* $\operatorname{Sgn} \delta_i = P_n(\sigma_i)$ *or $\delta_i = 0$, then $P_n(x)$ is an extremal polynomial of the segment $(\mu_k)_0^n$, since*

$$P_n(\bar{\mu}) = \sum_1^s |\delta_i| = N_n.$$

For brevity, we say in this case that the distribution of the polynomial is "suitable" for the segment $(\mu_k)_0^n$.

Thus we have obtain the following *criterion for a polynomial to be extremal.*

A necessary and sufficient condition that a reduced polynomial $P_n(x)$ is extremal for a given segment-functional $(\mu_k)_0^n$ is that if $(\overset{\pm}{\sigma}_i)_1^s$ is the distribution of $P_n(x)$, the system (23) *with $C_i = \sigma_i$ is consistent and that the solutions have signs given by*

$$\operatorname{Sgn} \Delta_k = P_n(\sigma_k), \quad \textit{or} \quad \Delta_k = 0 \ \textit{(but not all are zero)}.$$

We shall investigate some properties of the extremal polynomials of a given non-absolutely-monotonic segment $(\mu_i)_0^n$. Giving a segment is theoretically equivalent to giving the moment sequence $\mu_0, \cdots, \mu_n, \overset{*}{\mu}_{n+1}, \cdots, \overset{*}{\mu}_p, \cdots$, its best extension to infinity. This extension has an extremal polynomial of degree at most n. We shall refer to all such polynomials as extremal polynomials of the original segment. In the set of extremal polynomials there is a polynomial $Q_{\underline{m}}(x)$ of lowest degree $m \leqq n$ (possibly not unique); this is called a *principal polynomial.* We also note that if the segment $(\mu_k)_0^n$ is extended by one term in the best possible way, becoming $\mu_0, \cdots, \mu_n, \overset{*}{\mu}_{n+1}$, then an extremal polynomial $Q_n(x)$ for the original segment is also extremal for $(\mu_i)_1^{n+1}$, and if the norm of the given segment is $N = \sum \delta_i' + \sum |\delta_i''|$ then the norm of the new segment is $\sum \sigma_i' \delta_i' + \sum \sigma_i'' |\delta_i''|$.

Here $(\overset{\pm}{\sigma}_i)$ is the faithful distribution of $(\mu_k)_0^n$ and $Q_n(\sigma_i') = +1$, $Q_n(\sigma_i'') = -1$; $\delta_i' > 0$ and $\delta_i'' < 0$ are the corresponding weights.

Finally, if $(\overset{\pm}{\sigma}_i)_1^s$ is the faithful representation of a given $(\mu_k)_0^n$, or the distribution of a given reduced $P(x)$, then the polynomial $R_s(x) = \prod_1^s (x - \sigma_i)$ is called the resolvent of this segment or of this polynomial.

If $Q_{\underline{m}}(x)$ is a principal polynomial of the segment $(\mu_i)_0^n$ and $m < n$, the segment is extended in the best way starting from $\mu_{m+1} = \overset{*}{\mu}_{m+1}$; consequently the segment can be replaced by the truncated segment $(\mu_i)_0^m$. Hence we shall suppose that the principal polynomials of the segment $(\mu_k)_0^n$ are precisely of degree n ($\mu_n \neq \overset{*}{\mu}_n$); we call such a segment irreducible, and we call the number n the degree of the segment.

THEOREM 14. *If $Q(x)$ is an extremal polynomial of the segment $(\mu_i)_0^n$, every other extremal polynomial is of the form*

$$L(x) = Q(x) + \phi(x) R_s^2(x), \tag{25}$$

where $\phi(x)$ is a polynomial that ensures that $L(x)$ is reduced, and $R_s^2(x)$

is the "squared resolvent" of the segment, which has one of the following four forms (σ_i are the nodes of the segment):

1. $$R_s^2(x) = \prod_1^s (x - \sigma_i)^2, \quad \textit{if } 0 < \sigma_1 < \cdots < \sigma_s < 1.$$

2. $$R_s^2(x) = x\prod_2^s (x - \sigma_i)^2, \quad \textit{if } 0 = \sigma_1 < \cdots < \sigma_s < 1.$$

3. $$R_s^2(x) = (1 - x)\prod_1^{s-1} (x - \sigma_i)^2, \quad \textit{if } 0 < \sigma_1 < \cdots < \sigma_s = 1.$$

4. $$R_s^2(x) = x(1 - x)\prod_2^{s-1} (x - \sigma_i)^2, \quad \textit{if } 0 = \sigma_1 < \cdots < \sigma_s = 1.$$

In fact, since any two functions $f_1(x)$ and $f_2(x)$ are connected by a unique relation of the form $f_2(x) = f_1(x) + \Phi(x)$, we may put $L(x) = Q(x) + \Phi(x)$; by hypothesis $L(\sigma_i) = Q(\sigma_i) = \pm 1$, i.e. $\Phi(\sigma_i) = 0$; furthermore, Q and L have extrema at the points (σ) that are inside $[0,1]$, i.e. $Q'(\sigma_i) = 0$ and $L'(\sigma_i) = 0$, i.e. $\Phi'(\sigma_i) = 0$. It follows that $\Phi(x)$ is a multiple of $R_s^2(x)$, and (25) is established. For $\phi(x)$ we can take any polynomial such that $L(x)$ is reduced; we have $\max_{[0,1]}|L(x)| = 1$.

We divide the segments $(\mu_i)_0^n$ into two classes: the first class contains those for which the number s of nodes in the faithful distribution satisfies the condition that n is not less than the degree of the "squared resolvent"; the second class consists of the others, for which $s > \frac{1}{2}n + 1$ (with strict inequality to consolidate the four possible forms).

Considering the separate forms given by (25), we have 1) $n < 2s$, i.e. $s > n/2$; 2) and 3), $n < 2s - 1$, i.e. $s > \frac{1}{2}(n + 1)$; 4) $n < 2s - 2$, i.e. $s > \frac{1}{2}n + 1$.

COROLLARY 1. *For every segment of class* II *an extremal polynomial of degree $\leq n$ is unique (also principal).*

In fact, if $Q_m(x)$ satisfies $m \leq n$ then by (25) every other extremal polynomial has degree greater than n.

COROLLARY 2. *Let $(\mu_i)_0^n$ be a segment of class* II *with principal polynomial $Q_n(x)$ and with s faithful nodes ($\frac{1}{2}n + 1 \leq s \leq n + 1$). We extend the segment in the best way by one term, two terms, etc., as long as the resulting segments remain of class* II. *These segments $(\mu_i)_0^{n+1}, (\mu_i)_0^{n+2}, \cdots, (\mu_i)_0^p$ do not have extremal polynomials other than $Q_m(x)$ (of degree $m \leq n$).*

As we shall see later, a segment of class I can also have a unique principal polynomial, but this does not necessarily occur.

We now give another proof of the corollary of Theorem 14 on the uniqueness of the principal polynomial of a segment of class II; this corollary is fundamental for what follows. We had (p. 17) the following form for extremal polynomials $Q_n(x)$:

$$Q_n(x) \equiv 1 - \prod_1^{k_1} (x - \sigma_i')^2 \phi_{n-2k_2}(x) \equiv -1 + \prod_1^{k_2} (x - \sigma_i'')^2 \psi_{n-2k_1}(x)$$

(in the case of interior nodes). Here $Q_n(\sigma_i') = +1$; $Q_n(\sigma_i'') = -1$; $\phi(x) \geqq 0$; $\psi(x) \geqq 0$ for $0 \leqq x \leqq 1$; $k_1 + k_2 = s$, the total number of nodes of the segment. Then

$$\prod_1^{k_1} (x - \sigma_i')^2 \phi(x) + \prod_1^{k_2} (x - \sigma_i'')^2 \psi(x) \equiv 2, \tag{26}$$

but for a segment of class II (with interior nodes!) $n < 2s = 2k_1 + 2k_2$.

Consequently $n - 2k_1 < 2k_2$ and $n - 2k_2 < 2k_1$.

Therefore $\phi(x)$ and $\psi(x)$ in (26) are uniquely determined by the Euclidean algorithm, and since a principal polynomial must be of degree at most n, it will be found by this process, i.e. we necessarily obtain $\phi(x) \geqq 0$ and $\psi(x) \geqq 0$. Hence the uniqueness of the principal polynomial again follows.

The proof can be carried out in the same way when there are nodes at an endpoint of the interval.

Theorem 15 [7]. *If the segment $(\mu_i)_0^n$ has, besides the principal polynomial $Q_{\underline{n}}(x)$, an extremal polynomial $Q_{\underline{p}}(x)$ $(p \geqq n)$, then it has an infinite set of extremal polynomials of degree p and of arbitrary degree greater than p.*

In fact, the first statement is established by the weighted mean

$$\frac{\alpha Q_{\underline{n}}(x) + \beta Q_{\underline{p}}(x)}{\alpha + \beta}; \quad (\alpha > 0;\ \beta > 0),$$

since such a polynomial is also extremal.

In addition, by Theorem 14, $Q_{\underline{p}}(x) = Q_{\underline{n}}(x) + \phi(x) R_s^2(x)$ for any of the four forms, but then $Q_{\underline{n}}(x) + \psi(x)\phi(x) R_s^2(x)$ is also extremal when $\psi(x)$ is any polynomial for which $0 \leqq \psi(x) \leqq 1$ when $0 \leqq x \leqq 1$; for example, $\psi(x) = x$. This completes the proof.

THEOREM 16. *If* $(\mu_i)_0^n$ *is an irreducible segment such that its principal polynomial is not unique, then among its principal polynomials* $\{Q_n(x)\}$ *there is a unique one with leading coefficient* $q_{n(\max)}$ *and a unique one with leading coefficient* $q_{n(\min)}$.

In fact, the set $\{q_n\}$ for all (reduced) principal polynomials is bounded by the norm of the segment $(\mu_i)_0^n = 0_0, 0_1, \cdots, 0_{n-1}, 1_n$. Suppose that besides $Q_{\underline{n}}(x)$ with coefficient $q_{n(\max)}$ there is another polynomial $q_{\underline{n}}^{(1)}(x)$ with the same coefficient; then

$$Q_n^{(1)}(x) = Q_n(x) + \phi(x) R_s^2(x),$$

where the polynomial $\phi(x) R_s^2(x)$ has degree $k < n$; let the leading coefficient of this polynomial be p_k; if $\operatorname{Sgn} p_k = \operatorname{Sgn} q_{n(\max)}$, take $\psi(x) = x^{n-k}$; if $\operatorname{Sgn} p_k = -\operatorname{Sgn} q_{n(\min)}$, take $\psi(x) = x^{n-k-1}(1-x)$.

Then $Q_{\underline{n}}(x) + \psi(x)\phi(x) R_s^2(x)$ is an extremal polynomial of degree n with leading coefficient larger than $q_{n(\max)}$, which is impossible. A similar discussion applies to $q_{n(\min)}$. The theorem remains valid also for a reducible segment $(\mu_i)_0^n$ if there is an extremal polynomial of degree precisely n.

COROLLARY. *If* $q_{n(\max)} = q_{n(\min)}$ *for all principal polynomials, i.e. if the leading coefficient has only one value, the same is true for all other coefficients, i.e.* $Q_{\underline{n}}(x)$ *is unique.*

We now extend the concept of classes determined by segments or polynomials. Suppose that $\{Q_n(x)\}$ is the set of reduced polynomials of degree at most n. Let $(\overset{\pm}{\sigma}_i)_1^s$ be the distribution of a polynomial $Q_{\underline{m}}(x)$. We say that $Q_{\underline{m}}(x)$ is a polynomial of class II if $s > \frac{1}{2}n + 1$; otherwise $Q_{\underline{m}}(x)$ is of class I. It is clear that if $Q_n(x)$ is an extremal polynomial of the segment $(\mu_i)_0^n$ and this segment is of class II with s_1 nodes, then $Q_n(x)$ is a polynomial of class II, and $s \geqq s_1$.

It will be shown later that polynomials of class I are not needed for solving any of the problems that we shall consider.

§4. Čebyšev polynomials in the critical intervals [7]

In the preceding discussion we deliberately avoided reference to the Čebyšev polynomial $T_n(x) = \operatorname{Cos} n \arccos(2x-1)$, a reduced polynomial of class II on $[0,1]$. Supposing it is unknown, we begin by proving its existence and giving a method for its construction, using only the simplest segment-functional. By the properties we have established for a principal polynomial $Q_n(x)$ of a non-absolutely-monotonic segment $(\mu_i)_0^n$, we must have

$$Q_n(\bar{\mu}) = \sum_{k=0}^{n} q_k\mu_k = \sum_{1}^{s} \delta_i Q_n(\sigma_i) = \sum_{1}^{s} |\delta_i| = N,$$

where $(\delta_i)_1^s$ are defined by the system

$$\sum_{i=1}^{s} \delta_i \sigma_i^k = \mu_k \quad (k = 0, 1, \cdots, n). \tag{27}$$

Here $(\overset{\pm}{\sigma_i})_1^s$ is the faithful distribution of the segment, and $Q_n(\overset{+}{\sigma_i}) = +1$ and $Q_n(\overset{-}{\sigma_i}) = -1$. Consequently, the largest possible number of such points is $s = n + 1$, and then 0 and 1 must be nodes, since points (σ) inside $[0, 1]$ are zeros of $Q_n'(x)$. Thus when $s < n + 1$ the system (27) is overdetermined.

It is also quite obvious that when $s = n + 1$ the system (27) can be solved with alternating signs with respect to the (δ_i), i.e. two successive numbers σ have opposite signs, since otherwise the polynomial $Q_n'(x)$ would have more than $n - 1$ roots.

EXAMPLE 16. Put

$$(\mu_i)_0^n = 0_0, 0_1, \cdots, 0_{n-1}, 1_n \tag{28}$$

(here the subscripts indicate the indices of the terms). This segment is clearly non-absolutely-monotonic and irreducible. It is easily shown that it has $n + 1$ faithful nodes.

In fact, substituting these (μ_i) into (27) and taking the first n equations, we obtain a homogeneous system for (δ_i); if we suppose that $s < n + 1$ we find that all $\delta_i = 0$, which is impossible. Thus our segment is of class II and its extremal polynomial $Y_n(x)$, which certainly exists, is unique for the segment and has $n + 1$ nodes with alternating signs at these nodes. We remark that in general the polynomials $\pm Y_n(x)$ are the only reduced polynomials with $s = n + 1$ nodes. In fact, let $L_n(x)$ be another such polynomial with distribution $(\overset{\pm}{\rho_i})_0^n$, where, by what we have proved, the signs associated with the ρ_i must alternate, and 0 and 1 are included among the ρ's. Then one of the polynomials $\pm L_n(x)$ would also satisfy the criterion for being extremal for the segment $(\mu_i)_0^n$ defined by (28) (p. 36), which would contradict the uniqueness theorem (p. 37).

Some properties of the polynomials $Y_n(x)$ follow immediately from their defining segment-functional (28). We have $(\mu_{0,i})_{i=0}^{n} = 0_0, 0_1, \cdots, 0_{n-1}, (-1)^n$, and its extremal polynomial is $Y_n(1-x)$ (p. 19). Hence $Y_n(1-x) = (-1)^n Y_n(x)$ (symmetry).

To obtain $Y_n(x)$ explicitly we start with the case $n = 2$. By what we have proved, $Y_2(0) = Y_2(1) = 1$, and by symmetry (for $s = 3$) we have $Y_2(\frac{1}{2}) = -1$; consequently $Y_2(x) = 8x^2 - 8x + 1$. But

$$Y_2\left(\frac{1 + Y_2(x)}{2}\right)$$

is a reduced polynomial of degree 4 and has five points of deviation on $[0,1]$ with alternating sign; consequently

$$Y_2\left(\frac{1 + Y_2(x)}{2}\right) = Y_4(x);$$

by the same argument,

$$Y_8(x) = Y_2\left(\frac{1 + Y_4(x)}{2}\right),$$

and generally

$$Y_{2^n}(x) = Y_2\left(\frac{1 + Y_{2^{n-1}}(x)}{2}\right). \tag{29}$$

To find $Y_n(x)$ we formally replace 2^n by n in (29) and show that we obtain the required polynomial. Let $x = \frac{1}{2}(1 + \cos\phi)$; we have $Y_2(x) = \cos 2\phi$,

$$Y_4(x) = Y_2\left(\frac{1 + \cos 2\phi}{2}\right) = \cos 4\phi.$$

Putting $Y_{2^{n-1}}(x) = \cos 2^{n-1}\phi$, we have

$$Y_{2^n}(x) = Y_2\left(\frac{1 + \cos 2^{n-1}\phi}{2}\right) = \cos 2^n\phi,$$

i.e. $Y_n(x) = \cos n\phi$. On $0 \leqq \phi \leqq \pi$ this has $n+1$ points of deviation (± 1) and is an algebraic polynomial if we put $\phi = \arccos(2x-1)$, since $\cos\phi = 2x - 1$, $\cos 2\phi = 2(2x-1)^2 - 1$, and for $n \geqq 3$

$$\cos n\phi = \cos(n-2)\phi + 2\cos(n-1)\phi\cos\phi.$$

We have $Y_n(x) = T_n(x)$ (the Čebyšev polynomial). The distribution of a Čebyšev polynomial will be denoted by $(\overset{\pm}{\tau}_i)_0^n$, where $\tau_i = \sin^2(i\pi/2n)$

$(i = 0, 1, \cdots, n)$, and the node $\tau_n = 1$ carries a $+$ sign. In expanded form,

$$T_n(x) = 2^{2n-1}x^n - n2^{2n-2}x^{n-1} + \cdots$$
$$\cdots + (-1)^k \frac{n(2n-k-1)\cdots(2n-2k+1)}{k!} 2^{2n-2k}x^{n-k} + \cdots$$
$$\cdots + (-1)^{n-1}2nx^2 + (-1)^n.$$

It is easily seen that the segment $0_0, 0_1, \cdots, 0_{n-1}, -1_n$ has $-T_n(x)$ as its (unique) principal polynomial.

EXAMPLE 17. $(\mu_i)_0^n = 0_0, \cdots, 0_{k-1}, \pm 1_k, 0_{k+1}, \cdots, 0_n$ $(k > 0)$. It is easily verified that when $\mu_k = +1$ the (unique) principal polynomial is $(-1)^{n-k}T_n(x)$, and when $\mu_k = -1$ the corresponding polynomial is $(-1)^{n-k+1}T_n(x)$ $(k > 0)$. In fact, solving the system (27) for this segment with nodes $(\tau_i)_0^n$ we obtain δ_i of alternating sign, which establishes our assertion in view of the theorem used in the preceding example.

We note a (known) corollary. Among all polynomials $\{P_n(x)\}$ for which $\max_{[0,1]}|P_n(x)| = 1$, the polynomials $\pm T_n(x) = \pm\sum_0^n t_i x^i$ are those, and the only ones, that provide the absolute maximum for each coefficient, i.e. $|p_k| \leqq |t_k|$ $(k \geqq 0)$. In fact, as we have shown, $|T_n(\bar{\mu})| = N = |t_k|$.

The existence of the largest possible number of nodes for the polynomials $\pm T_n(x)$ naturally puts these polynomials in a special position with respect to the rest of the polynomials of the set $\{Q_n(x)\}$. We now make this more precise [7].

THEOREM 17. *For any segment* $(\mu_i)_0^n$, *if* $|T_n(\bar{\mu})| < N$ *we can find*

1) *a number* $h_1 > 0$ *so large that* $+T_n(x)$ *is extremal for the segment*

$$\mu_0, \cdots, \mu_i, \cdots, \mu_{n-1}, \mu_n + h_1; \tag{30}$$

2) *a number* $h_2 > 0$ *so large that* $-T_n(x)$ *is extremal for the segment*

$$\mu_0, \cdots, \mu_i, \cdots, \mu_{n-1}, \mu_n - h_2. \tag{30'}$$

Let $0 = \tau_0 < \tau_1 < \cdots < \tau_{n-1} < \tau_n = 1$ be the nodes of $T_n(x)$, so that $T_n(\tau_i) = (-1)^{n-i}$. We solve (27) for $(\mu_i)_0^n$, with $\sigma = \tau$.

We introduce the following notations:

$V_{n+1}(\tau_0, \cdots, \tau_n)$ is the Vandermonde determinant of $(\tau_i)_0^n$.

$V_n^{(i)}(x)$ is the polynomial obtained by replacing the column of V_{n+1} that contains powers of τ_i by a column of powers of x.

$V_{n,i}$ is the minor of τ_i^n in the determinant V_{n+1}.

Then (27) has the solution

$$\delta_i = \frac{V_n^{(i)}(\bar{\mu})}{V_{n+1}(\tau_0, \cdots, \tau_n)} \quad (i = 0, 1, \cdots, n).$$

If we replace μ_n by $\mu_n + h$ and δ_i by δ_i' in (27), we obtain

$$\delta_i' = \frac{V_n^{(i)}(\bar{\mu}) + (-1)^{n-i} V_{n,i} h}{V_{n+1}(\tau_0, \cdots, \tau_n)}.$$

In the last formula the denominator and $V_{n,i}$ are positive, and hence h can be taken so large that $\operatorname{Sgn} \delta_i' = (-1)^{n-i}$ for $i = 0, 1, \cdots, n$. Then the signs of the δ_i' will alternate, starting with $+$ for $i = n$, and $\operatorname{Sgn} \delta_i' = \operatorname{Sgn} T_n(\tau_i)$. Therefore $+ T_n(x)$ is an extremal polynomial for the segment (30).

Similarly we can verify the second assertion for the segment (30′).

Corollary 1. *If $+ T_n(x)$ is an extremal polynomial for the segment $(\mu_i)_0^n$, it remains extremal if μ_n is increased; if $- T_n(x)$ is extremal, it remains extremal if μ_n is decreased.*

Corollary 2. *There are two numbers $\mu_n' \leqq \mu_n''$ with the following properties: the norm of the segment $\mu_0, \cdots, \mu_{n-1}, \mu_n'' + h$ is attained by $+ T_n(x)$ when $h \geqq 0$, and not when $h < 0$; and similarly for $\mu_0, \cdots, \mu_{n-1}, \mu_n' - h$ and $- T_n(x)$.*

Hence $\pm T_n(x)$ cannot be extremal for $(\mu_i)_0^n$ if $\mu_n' < \mu_n < \mu_n''$; we call the open interval (μ_n', μ_n'') the *critical interval* for μ_n.

We have now obtained a property of the polynomials $\pm T_n(x)$ which we call their *extendibility* outside the critical interval for μ_n (extendibility of the property of being extremal), and nonextendibility inside the interval.

To determine the endpoints μ_n' and μ_n'' of the critical interval we may proceed as follows (first method).

Solve the system (27) with $(\sigma_i)_1^s$ replaced by $(\tau_i)_0^n$ and μ_n replaced by $\mu_n + h$, thus obtaining δ_i' $(i = 0, 1, \cdots, n)$; then solve separately the $n + 1$ equations $\delta_i' = 0$ for h.

Among all the values obtained for h, select the largest, say h'', and the smallest, say h'. Then $\mu_n' = \mu_n + h'$ and $\mu_n'' = \mu_n + h''$, i.e. μ_n' and μ_n'' are uniquely determined. In particular cases the critical interval may reduce to a point (examples later).

Theorem 17 can be generalized to any term μ_k $(k > 0)$; in fact, by the same method we can prove Theorem 18.

Theorem 18. *Let the segment* $\mu_0, \mu_1, \cdots, \mu_{k-1}, \mu_k + h, \mu_{k+1}, \cdots, \mu_n$ *be given, with* $k > 0$, *and suppose that* $|T_n(\bar{\mu})| < N$ *when* $h = 0$. *For sufficiently large* $h > 0$, *the polynomial* $+T_n(x)$ *is extremal for the segment if* $k \equiv n \pmod 2$, *and* $-T_n(x)$ *is extremal if* $k + 1 \equiv n \pmod 2$.

(*Similarly for* $h < 0$ *and sufficiently large* $|h|$.)

In fact, as in Theorem 17 we have the formulas

$$\delta_i' = \frac{V_n^{(i)}(\bar{\mu}) + (-1)^{k-i} V_{k,i} h}{V_{n+1}(\tau_0, \cdots, \tau_n)}$$

with an analogous result for the sign of δ_i' since $V_{k,i} > 0$.

Thus each element μ_k $(k > 0)$ has a "critical interval" when the other elements are held fixed.

Corollary 1. *If one element of the segment* $(\mu_i)_0^n$ *lies outside (inside) its critical interval, then every other element lies outside (inside) its critical interval. Thus all* $(\mu_i)_1^n$ *lie simultaneously at the ends of their critical intervals, and alternately at the right-hand and left-hand ends, if we start from* $i = n$.

Corollary 2. *A necessary and sufficient condition that the segment* $(\mu_i)_0^n$ *has either* $+T_n(x)$ *or* $-T_n(x)$ *as an extremal polynomial is that one of the numbers* μ_i $(i > 0)$ *lies outside or on the boundary of its critical interval.*

The only exception to these rules occurs for μ_0, which in general does not possess a critical interval. In fact, if we replace μ_0 by $\mu_0 + h$ and express the segment by means of the nodes $(\tau_i)_0^n$ we obtain (by the generalization of Theorem 17) $\delta_k' = \delta_k$ $(k = 1, 2, \cdots, n)$ and $\delta_0' = \delta_0 + h$ (since $V_{0,i} = 0$ for $i > 0$).

Thus it may happen that none of the polynomials $\pm T_n(x)$ is extremal for any h. If $Q_n(x) = \sum_0^n q_k x^k$ is an extremal polynomial for $(\mu_i)_0^n$ and if $q_0 = +1$, then $Q_n(x)$ remains extremal if μ_0 is replaced by $\mu_0 + h$, where h is any positive number; if $q_0 = -1$, then $Q_n(x)$ remains extremal for $\mu_0 - h$, where $h > 0$. Therefore when $|q_0| = 1$ the polynomial $Q_0(x)$ is "extendible" for μ_0. Finally, if $|q_0| < 1$, the point 0 is not a node of the polynomial and, as (27) shows, the polynomial is not extendible for μ_0.

Corollary 3. *If the segment* $(\mu_i)_0^n$ *is such that its element* μ_k $(k > 0)$ *lies in its critical interval (and consequently all its elements do so), the number* s *of nodes of an extremal polynomial* $Q_n(x)$, *and therefore the number* s_1 $(\leqq s)$ *of faithful nodes, is less than* $n + 1$; *the system* (27) *is overdetermined; consequently* (δ_i) *are determined by any* s_1 *equations of* *the system including the first one; then the remaining $n+1-s_1$ equations are also satisfied. Therefore, a given polynomial from the set $\{Q_n(x)\}$ (other than $\pm T_n(x)$) is extremal at most for one value of μ_k in the critical interval (at one point of the interval).*

Thus $Q_n(x)$, an extremal polynomial for $(\mu_i)_0^n$ with $\mu_k' < \mu_k < \mu_k''$, is not extendible as an extremal polynomial for any element μ_k $(k > 0)$.

To determine the endpoints of the critical intervals for the elements (μ_i), it is not necessary to solve (27); one can proceed in a different way (second method), by using the resolvent of the polynomials $\pm T_n(x)$, i.e.

$$R_{n+1}(x) = \frac{x(x-1)T_n(x)}{n2^{2n-1}} = \prod_0^n (x - \tau_i).$$

Form the $n+1$ polynomials

$$R_n^{(i)}(x) = \frac{R_{n+1}(x)}{x - \tau_i} \qquad (i = 0, 1, \cdots, n)$$

and suppose that $+T_n(x)$ is an extremal polynomial for $(\mu_i)_0^n$; then, since

$$R_n^{(i)}(\bar{\mu}) = \sum_{k=0}^n \delta_k R_n^{(i)}(\tau_k) = \delta_i R_n^{(i)}(\tau_i),$$

we have

$$\text{(31)} \qquad \delta_i = \frac{R_n^{(i)}(\bar{\mu})}{R_n^{(i)}(\tau_i)} \quad \text{or} \quad \delta_i = \frac{R_n^{(i)}(\bar{\mu})}{R_{n+1}'(\tau_i)} \qquad (i = 0, 1, \cdots, n),$$

since

$$\lim_{x \to \tau_i} R_n^{(i)}(x) = \lim_{x \to r_i} \frac{R_{n+1}(x)}{x - r_i} = R_{n+1}'(\tau_i).$$

Here, by hypothesis, the (δ_i) have alternating signs, starting with $+$ for $i = n$ (or zero).

It is easy to see from the graph of $R_{n+1}(x)$ that the signs of $R_{n+1}'(\tau_i)$ alternate for $i = 0, 1, \cdots, n$, and in fact $\operatorname{Sgn} R_{n+1}'(\tau_i) = (-1)^{n-i}$. Consequently a necessary and sufficient condition that $+T_n(\bar{\mu}) = N$ is that all the numbers $R_n^{(i)}(\bar{\mu})$ are nonnegative. If one $\delta_i = 0$ then all the μ's lie at endpoints of their critical intervals. (Similarly when $N = -T_n(\bar{\mu})$ we must have $\delta_i \leqq 0$.)

We note that $R_n^{(i)}(x)$ does not depend on (μ), and this is the advantage of the second method. To find the endpoints of the critical interval for an element μ_k suppose for definiteness that $k \equiv n \pmod 2$; putting

$\mu_k = x$ $(k > 0)$, we form the linear equations $R_n^{(i)}(\bar{\mu}) = 0$ $(i = 0, 1, \cdots, n)$; solving them for x, we obtain $n+1$ roots $x^{(0)}, x^{(1)}, \cdots, x^{(n)}$. Let $x_{\max}$ and $x_{\min}$ be the largest and smallest roots. Then $x_{\max} = \mu_k''$; $x_{\min} = \mu_k'$. In fact, $R_n^{(i)}(\bar{\mu}) = f^{(i)}(x)$ is a linear function whose sign changes only at $x = x^{(i)}$. Since the zeros of all the polynomials $R_n^{(i)}(x) = \sum_{p=0}^{n} r_p^{(i)} x^p$ are positive, we have $(r_p^{(i)})_{p=0}^{n}$ of alternating sign ($i =$ const.) and $(r_p^{(i)})_{i=0}^{n}$ of fixed sign ($p =$ const.). If we take $\mu_k \geqq x_{\max}$ and put $\mu_k = x^{(i)} + h^{(i)}$ $(h^{(i)} \geqq 0)$, we have $R_n^{(i)}(\bar{\mu}) = r_k^{(i)} h^{(i)}$, i.e. the $R_n^{(i)}(\bar{\mu})$ $(i = 0, 1, \cdots, n)$ are of fixed sign (or zero); and since $r_k^{(n)} > 0$ by hypothesis, $+\,T_n(x)$ is extremal for all $\mu_k \geqq x_{\max}$, and this is an endpoint. (Similarly for $x_{\min}$.) For k and n of opposite parity, $x_{\min} = \mu_k''$ and $x_{\max} = \mu_k'$.

EXAMPLE 18 (SECOND METHOD). For the segment $(\mu_i)_0^n = 0_0, 0_1, \cdots, 0_{n-2}, 1_{n-1}, x_n$, find μ_n' and μ_n''.

Form

$$x(x-1)\,T_n'(x) = x(x-1)\,[n2^{2n-1}x^{n-1} - n(n-1)2^{2n-2}x^{n-2} + \cdots]$$

$$= n2^{2n-1}x^{n+1} - [n^2 2^{2n-2} + n2^{2n-2}]x^n + \cdots;$$

$$n2^{2n-1}R_n^{(i)}(x) = n2^{2n-1}x^n - [n^2 2^{2n-2} + n2^{2n-2} - n2^{2n-1}\tau_i]x^{n-1} + \cdots;$$

$$n2^{2n-1}R_n^{(i)}(\bar{\mu}) = n2^{2n-1}x - n2^{2n-2}(n+1-2\tau_i);$$

$$x_{\max} = \frac{n+1}{2} \text{ (when } \tau_i = 0); \quad x_{\min} = \frac{n-1}{2} \text{ (when } \tau_i = 1).$$

To find μ_n^*, the number that extends the segment in the best (unique!) way, we note that in our example $\mu_{0,n} = (-1)^{n-1}(n-x)$ and $\mu_{0,k} = 0$ for $k \leqq n-2$; $\mu_{0,n-1} = (-1)^{n-1}$. Therefore to μ_n^+ there corresponds the number $\mu_{0,n}^*$, the best extension of the segment $(\mu_{0,i})_0^{n-1}$. Consequently we must have $|\mu_{0,n}^*| = \mu_n^*$, i.e. $\mu_n^* = n/2$.

As a corollary of this calculation we find that when $n \geqq 3$ the point $x = 1$ is not inside its critical interval $(1 \leqq \frac{1}{2}(n-1))$. Consequently the segment $0_0, \cdots, 0_{n-2}, 1_{n-1}, 1_n$ has $-\,T_n(x)$ as extremal polynomial, i.e. the sum of the two leading coefficients of any polynomial $P_n(x)$ satisfies the inequality

$$|p_n + p_{n-1}| \leqq 2^{2n-2}(n-2) \cdot \max_{[0,1]} |P_n(x)|.$$

EXAMPLE 19 (FIRST METHOD). Find the critical intervals for μ_n and ν_n for the following two segments:

$$(\mu_i)_0^n = 1_0, 0_1, \cdots, 0_{n-1}, x, \quad \text{and} \quad (\nu_i)_0^n = 1_0, 1_1, \cdots, 1_{n-1}, y.$$

We note first that

$$\mu_i = \nu_{0,i} \quad \text{for } i = 0, 1, \cdots, n-1$$
$$\mu_n = x = (-1)^{n-1}(1-y).$$

Hence if $x = \mu_n''$ then $y = \nu_n''$ when n is even, and $y = \nu_n' = 1 - \mu_n''$ when n is odd (similarly for $x = \mu_n'$).

Expressing $(\mu_i)_0^n$ in terms of the nodes $(\tau_i)_0^n$ by means of (27), we obtain

$$\delta_k = \frac{(-1)^{n-k} x V_{n,k}(\tau_0, \cdots, \tau_n)}{V_{n+1}(\tau_0, \cdots, \tau_k)} \quad (k \neq 0);$$

$$\delta_0 = \frac{\prod_{i \neq 0} \tau_i V_n^{(0)}(\tau_1, \cdots, \tau_n) + (-1)^n x V_{n,0}(\tau_1, \cdots, \tau_n)}{V_{n+1}(\tau_0, \cdots, \tau_n)}.$$

Hence, by putting $\delta = 0$, we obtain

$$\mu_n'' = 0 \quad \text{and} \quad \mu_n' = -\prod \tau_i \quad (i \neq 0) \text{ when } n \text{ is even;}$$

$$\mu_n' = 0 \quad \text{and} \quad \mu_n' = \prod \tau_i \quad (i \neq 0) \text{ when } n \text{ is odd.}$$

In both cases $\mu_n^* = 0$, i.e. $V_n^* = 1$.

Example 20 (first method). For the segment $(\mu_i)_0^n = 1, \rho, \rho^2, \cdots, \rho^{n-1}, \rho^n$, with $\rho > 1$, find the critical intervals for each element μ_m $(m > 0)$.

We replace ρ^m by $\rho^m + \theta$ and then express the segment in terms of the nodes $(\tau_i)_0^n$. From (27) we obtain

$$\delta_k = (-1)^{n-k} \frac{\prod_{i \neq k} (\rho - \tau_i) + (-1)^{n-m}\theta \cdot s_{n-m}^{(k)}}{\prod_{i \neq k} |\tau_k - \tau_i|} \quad (k = 0, 1, \cdots, n).$$

Here $s_{n-m}^{(k)}$ is the sum of the products containing $n - m$ factors other than τ_k chosen from (τ_i).

Putting $\delta_k = 0$, we have

$$\theta = (-1)^{k+m-1} \cdot \frac{\prod (\rho - \tau_i)}{s_{n-m}^{(k)}} \quad (i \neq k).$$

From this it is clear that

$$|\theta|_{\min} = \frac{\prod (\rho - \tau_i)}{s_{n-m}^{(0)}} \quad (i \neq 0) \quad \text{and} \quad |\theta|_{\max} = \frac{\prod (\rho - \tau_i)}{s_{n-m}^{(n)}} \quad (i \neq n).$$

Thus the endpoints of the critical interval for μ_m are:

when $n - m$ is even,

$$\mu'_m = \rho^m - \frac{\prod(\rho - \tau_i)}{s^{(n)}_{n-m}} \quad (i \neq n); \qquad \mu''_m = \rho^m - \frac{\prod(\rho - \tau_i)}{s^{(0)}_{n-m}} \quad (i \neq 0);$$

when $n - m$ is odd,

$$\mu'_m = \rho^m + \frac{\prod(\rho - \tau_i)}{s^{(0)}_{n-m}} \quad (i \neq 0); \qquad \mu''_m = \rho^m + \frac{\prod(\rho - \tau_i)}{s^{(n)}_{n-m}} \quad (i \neq n).$$

EXAMPLE 21. The critical interval for the element $\mu_m = x$ in the segment

$$(\mu_k)_0^n = 0_0, 0_1, \cdots, 0_{m-1}, x, 0_{m+1}, \cdots, 0_n$$

reduces to the point $\mu'_m = \mu''_m = 0$, as follows from Example 17, with the elements each multiplied by x.

EXAMPLE 22 (SECOND METHOD). $(\mu_i)_0^n = 1, \tau_k, \tau_k^2, \cdots, \tau_k^{n-1}, \tau_k^n$, where $T_n(\tau_k) = +1$. To find μ'_n and μ''_n.

Replace τ_k^n by $\tau_n^k + h$. The solution of (27) is

$$\delta'_i = \frac{R_n^{(i)}(\tau_k) + h}{R'_{n+1}(\tau_k)} \qquad (i = 0, 1, \cdots, n).$$

According to our rule, we have to find $h_{\max}$ and $h_{\min}$ among the numbers $h = -R_n^{(i)}(\tau_k)$ $(i = 0, 1, \cdots, n)$. But for $i \neq k$ we have $R_n^{(i)}(\tau_k) = 0$, and $R_n^{(k)}(\tau_k) = R'_{n+1}(\tau_k) > 0$.

Therefore $h_{\min} = -R_n^{(k)}(\tau_k)$, $h_{\max} = 0$; and consequently $\mu'_n = \tau_k^n - R'_{n+1}(\tau_k)$; $\mu''_n = \tau_k^n$. It is clear that $\mu''_n = \mu_n^*$ in this case, since the segment is absolutely monotonic.

EXAMPLE 23. Given an amorphous segment $\alpha_0, \alpha_1, \cdots, \alpha_{n-1}$, find the condition under which the critical interval (μ'_n, μ''_n) for the unknown element α_n coincides with the interval (α'_n, α''_n) of best extension.

For the polynomial $+T_n(x)$ with distribution $(\overset{\pm}{\tau}_i)_0^n$ put $(\overset{+}{\tau}_i) = (\tau'_i)$ and $(\overset{-}{\tau}_i) = (\tau''_i)$; then from the condition $\alpha''_n = \mu''_n$ we obtain that $\alpha_0, \alpha_1, \cdots, \alpha_{n-1}, \mu''_n$ $(*_1)$ has faithful nodes $(\overset{+}{\tau'_i})$ and $\alpha_0, \alpha_1, \cdots, \alpha_{n-1}, \mu'_n$ $(*_2)$ has faithful nodes $(\overset{+}{\tau''_i})$.

Therefore the numbers $(\alpha_k)_0^{n-1}$ have two nodal structures of the form $\alpha_k = \sum_i \delta'_i \cdot \tau_i'^k = \sum_i \delta''_i \tau_i''^k$. To find (δ'_i) and (δ''_i) we form the segment $0_0, 0_1, \cdots, 0_{n-1}$, A_n, with $A_n = \mu''_n - \mu'_n > 0$, of the termwise difference of the segments $(*_1)$ and $(*_2)$ and express it in terms of the nodes $(\tau_i)_0^n$ (Example 17). We obtain δ_i of alternating sign. The set of positive δ_i yields δ'_i, and the set of negative δ_i yields $-\delta''_i$.

CHAPTER II

PROPERTIES OF THE NORM: SEGMENTS OF THE FIRST AND SECOND CLASSES

§1. The norm as a function of the elements

It was shown in §3 of Chapter I that for a non-absolutely-monotonic segment $(\mu_i)_0^{n-1}$ there is a unique number μ_n^* that extends the segment in the best way, i.e. with preservation of its norm. It is clear that μ_n^* cannot lie outside the critical interval: $\mu_n' \leqq \mu_n^* \leqq \mu_n''$. This follows from the fact that the extremal polynomial of the segment $(\mu_i)_0^n$ with $\mu_n = \mu_n'' + h$ is $+ T_n(x)$, and so the norm $N(h)$ increases with h; when $\mu_n = \mu_n' - h$ $(h \geqq 0)$ the extremal polynomial is $- T_n(x)$ and the norm again increases with h. We call the point μ_n^* the *focus* of the critical interval for μ_n and generalize this concept to the other elements μ_k.

In the segment $(\mu_i)_0^n$ let μ_k be variable; we say that the segment is completed in the best way at the kth place if the smallest norm is obtained when $\mu_k = \mu_k^*$.

THEOREM 19 [7]. *If the segment $(\mu_i)_0^n$ [or $(-\mu_i)_0^n$] does not become absolutely monotonic for any value of μ_k there is a unique number μ_k^* which completes the segment in the best way; μ_k^* is called the focus of the element μ_k, and when $k > 0$ it is not outside the critical interval $[\mu_k', \mu_k'']$.*

It is evident that there is a number μ_k^* that minimizes the norm, since $N(\mu_k)$ is continuous (p. 21). In addition, the necessity of the condition $\mu_k' \leqq \mu_k^* \leqq \mu_k''$ (for $k > 0$) can be proved in the same way as for $k = n$. By hypothesis, $N(\mu_k) > |\mu_0|$ for all μ_k, and consequently the segment always has an extremal polynomial $(\not\equiv 1)$. Suppose that there are two numbers $\mu_k^{(1)}$ and $\mu_k^{(2)}$ that minimize the norm $N(\mu_k)$. Then the segments

$$\text{(I)} \qquad (\nu_i)_0^n = \mu_0, \cdots, \mu_{k-1}, \mu_k^{(1)}, \mu_{k+1}, \cdots, \mu_n;$$

$$\text{(II)} \qquad (\lambda_i)_0^n = \mu_0, \cdots, \mu_{k-1}, \mu_k^{(2)}, \mu_{k+1}, \cdots, \mu_n$$

have the same norm $N_{\min}$. Take any $\alpha \geqq 0$ and $\beta \geqq 0$ (α and β not both zero), and construct a third segment

$$\text{(III)} \qquad (\xi_i)_0^n = \mu_0, \mu_1, \cdots, \mu_{k-1}, \frac{\alpha\mu_k^{(1)} + \beta\mu_k^{(2)}}{\alpha + \beta}, \mu_{k+1}, \cdots, \mu_n.$$

This segment is obtained by multiplying (I) and (II) respectively by $\alpha/(\alpha+\beta)$ and $\beta/(\alpha+\beta)$ and adding them term by term.

Since the norm of the sum of two segments does not exceed the sum of their norms, the norm of (III) satisfies

$$N_1 \leqq \frac{\alpha N_{\min}}{\alpha+\beta} + \frac{\beta N_{\min}}{\alpha+\beta} = N_{\min}.$$

Consequently $N_1 = N_{\min}$. Since by suitable choice of α and β the number $(\alpha\mu_k^{(1)} + \beta\mu_k^{(2)})/(\alpha+\beta)$ can be any number in $[\mu_k^{(1)}, \mu_k^{(2)}]$, it follows that every number in this interval minimizes the norm.

Let $Q_n(x) = \sum_0^n q_i x^i$ be an extremal polynomial of (III) with a fixed μ_k, where $\mu_k^{(1)} < \mu_k < \mu_k^{(2)}$.

1. If $q_k > 0$, and we replace μ_k by $\mu_k + h \leqq \mu_k^{(2)}$ with $h > 0$, the norm is not increased; at the same time, $Q_n(\bar{\xi}) = N_{\min} + q_k \cdot h > N_{\min}$, which is impossible.

2. If $q_k < 0$, we replace μ_k by $\mu_k - h \geqq \mu_k^{(1)}$ with $h > 0$. We obtain $Q_n(\bar{\xi}) = N_{\min} + |q_k| h > N_{\min}$, which is also impossible.

3. If $q_k = 0$, this is true for every μ_k in the interval $(\mu_k^{(1)}, \mu_k^{(2)})$, since the other alternatives are impossible.

Suppose now that $k > 0$; then $Q_n(x)$ remains extremal for arbitrary μ_k and for the whole interval, since the norm does not decrease. Consequently $Q_n(x)$, which is not $\pm T_n(x)$, is extendible by one element inside the critical interval, and this is impossible (Corollary 3 of Theorem 18).

Thus $\mu_k^{(1)} = \mu_k^{(2)} = \mu_k^*$.

There remains the case $k = 0$ with $q_0 = 0$.

We note that when $\mu_0 = \mu_0^*$ the segment $(\mu_i)_0^n$ does not have $\sigma = 0$ as a faithful node; in fact, let $(\overset{\pm}{\sigma}_i)_1^s$ be the faithful nodes of the segment, and suppose that $\sigma_1 = 0$ with weight δ $(\neq 0)$; then the extremal polynomial $Q_n(x)$ yields $Q_n(0) = q_0 = \pm 1$, which is excluded by hypothesis [**3**]. Then (27) is overdetermined, and $Q_n(x)$ is not extendible for the element μ_0. This completes the proof of Theorem 19.

We shall call the interval (μ_k', μ_k^*) the left-hand part of the critical interval, and (μ_k^*, μ_k'') its right-hand part $(k > 0)$. In the segment $(\mu_i)_0^n$ let one element $\mu_k = \theta$ $(k > 0)$, where θ varies in the critical interval $\mu_k' \leqq \theta \leqq \mu_k''$.

THEOREM 20. *If $\mu_k = \theta$ $(k > 0)$ in a non-absolutely-monotonic segment $(\mu_i)_0^n$, then the norm $N(\theta)$ is a continuous function which decreases monotonically in the left-hand part of the critical interval and increases mono-*

tonically in the right-hand part. If $Q_n(x,\theta) = \sum_{i=0}^n q_i(\theta)x^i$ *is a principal polynomial of the segment* $(\mu_i)_0^n$ *(when it is not unique this polynomial can be selected arbitrarily), then* $q_k(\theta)$ *increases on the whole critical interval, and* $q_k < 0$ *in the left-hand part,* $q_k > 0$ *in the right-hand part.*

For the increment of the norm we have the inequalities

$$(32) \qquad q_k(\theta)h < N(\theta+h) - N(\theta) < q_k(\theta+h)\cdot h \quad (h \lessgtr 0).$$

Take $\theta = \mu_k^* + h$ $(h \lessgtr 0)$. Then

$$N(\mu_k^*) < N(\mu_k^* + h) = Q_n(\bar{\mu}^*, \theta) + q_k(\theta)h.$$

Since $Q_n(\bar{\mu}^*, \theta) < N(\mu_k^*)$, we must have $q_k(\theta) > 0$ for $h > 0$ and $q_k(\theta) < 0$ for $h < 0$.

Now put $\theta = \mu_k + h$, where $\mu_k \geqq \mu_k^*$ and $h > 0$; let the resulting segment be $(\nu_i)_0^n$ and let an extremal polynomial for it be $L_n(x) = \sum_0^n l_i x^i$.

We have

$$L_n(\bar{\nu}) = L_n(\bar{\mu}) + l_k h = N(\mu_k + h);$$

$$Q_n(\bar{\nu}) = Q_n(\bar{\mu}) + q_k h = N(\mu_k) + q_k h < N(\mu_k + h),$$

but $q_k > 0$; consequently $N(\mu_k + h) = L_n(\bar{\mu}) + l_k h > N(\mu_k) + q_k h$, but since $L_n(\bar{\mu}) < N(\mu_k)$ it follows that the norm increases and $q_k < l_k$ to the right of the focus. Similar reasoning shows that $N(\theta)$ decreases to the left of the focus and that q_k increases. It remains to establish (32).

Putting $\theta = \mu_k$, we have, in the same notation,

$$N(\mu_k) > L_n(\bar{\mu}) = N(\mu_k + h) - l_k h,$$

whence

$$N(\mu_k + h) - N(\mu_k) < l_k h = q_k(\theta + h)\cdot h;$$

$$N(\mu_k + h) > Q_n(\bar{\nu}) = N(\mu_k) + q_k h,$$

whence $N(\mu_k + h) - N(\mu_k) > q_k h$ and the inequality is established.

THEOREM 21. *The norm* $N(\mu_k)$ *of the segment* $(\mu_i)_0^n$ *has right-hand and left-hand derivatives at each point* μ_k $(k > 0)$ *of the critical interval* $\mu_k' \leqq \mu_k \leqq \mu_k''$, *and* $N_d'(\mu_k) = q_{k(\max)}$ *and* $N_s'(\mu_k) = q_{k(\min)}$. *The values* $q_{k(\max)}$ *and* $q_{k(\min)}$ *are analogous to* $q_{n(\max)}$ *and* $q_{n(\min)}$ *of Theorem* 16.

Let $\mu_k > \mu_k^*$. Let $Q_n^{(k)}(x) = \sum_0^n q_i(\mu_k + h)x^i$ be an extremal polynomial of the segment $(\nu_i)_0^n$ obtained from $(\mu_i)_0^n$ by replacing μ_k by $\mu_k + h$; let $h \geqq 0$ and consider the coefficient $q_k(\mu_k + h)$. By Theorem 20, $q_k(\mu_k + h)$ decreases as $h \to 0+$, remaining larger than $q_k(\mu_k)$.

Thus $\bar{q}_k = \lim_{h\to 0+} q_k(\mu_k + h)$ exists.

Consider the remaining coefficients $q_i(\mu_k + h)$, where $i \neq k$. As $h \to 0$ they are bounded: $|q_i(\mu_k + h)| < |t_i|$, where t_i is the corresponding coefficient of the Čebyšev polynomial $T_n(x)$. Hence $\overline{\lim}_{h\to 0+} q_i(\mu_k + h)$ and $\underline{\lim}\, q_i(\mu_k + h)$ exist for each $i \neq k$ [1].

We choose a sequence $h_p \to 0$ such that the $q_i(\mu_k + h_p)$ have limits $\lim_{p\to\infty} q_i(\mu_k + h_p) = \bar{q}_i$ $(i = 0, 1, \cdots, n)$; then

$$\lim_{h_p\to 0} Q_n^{(h_p)}(x) = \overline{Q}_n(x) = \sum_0^n \bar{q}_i x^i.$$

It is easy to show that this polynomial is an extremal polynomial for the segment $(\mu_i)_0^n$. In fact, 1) $\max_{[0,1]} |\overline{Q}_n(x)| = 1$, since $\max_{[0,1]} |Q_n^{(h)}(x)| = 1$ for arbitrary $h_p \to 0$; 2) $\overline{Q}_n(\bar{\mu}) = N$ since

$$\overline{Q}_n(\bar{\mu}) = \lim_{h_p\to 0} \overline{Q}_n^{(h)}(\bar{\nu}) = \lim_{h_p\to 0} N(\mu_k + h_p) = N(\mu_k)$$

by the continuity of the norm (Theorem 5). Thus there is an extremal polynomial for the segment $(\mu_i)_0^n$ for which each coefficient is the limit of the corresponding coefficient of extremal polynomials of the segment with μ_k replaced by $\mu_k + h_p$ and $h_p \to 0+$. Since the set of extremal polynomials $\{Q_n(x)\}$ of the segment $(\mu_i)_0^n$ (if there is not a unique extremal polynomial) contains a polynomial for which $q_k = q_{k(\max)}$, it follows that

$$\lim_{h\to 0+} q_k(\mu_k + h) = \bar{q}_k = q_{k(\max)}.$$

A similar discussion for $h < 0$ yields $\lim_{h\to 0-} q_k(\mu_k + h) = q_{k(\min)}$. By using inequality (32) we can sharpen Theorem 20; in fact, for the left-hand part of the critical interval

(33)
$$q_k(\mu_k + h) < \frac{\Delta N}{h} < q_{k(\min)} \text{ for } h < 0;$$
$$q_{k(\max)} < \frac{\Delta N}{h} < q_k(\mu_k + h) \text{ for } h > 0.$$

Consequently $q_{k(\max)} = N'_d(\mu_k)$ and $q_{k(\min)} = N'_s(\mu_k)$.

COROLLARY 1. *A necessary and sufficient condition for the uniqueness of q_k, the kth coefficient of a principal polynomial of the segment $(\mu_i)_0^n$, is the existence of $\partial N/\partial \mu_k$, and then $q_k = \partial N/\partial \mu_k$. In the special case when $\mu_k = \mu_k^*$ this condition becomes $(\partial N/\partial \mu_k)_{\mu_k = \mu_k^*} = 0$. For a segment of class*

I *the existence of* $\partial N/\partial\mu_n = q_n$ *is sufficient for the principal polynomial to be unique (corollary of Theorem* 16).

COROLLARY 2. *If for* $\mu_k = \Theta$ $(k > 0)$ *the segment* $(\mu_i)_0^n$ *is such that the principal polynomial* $Q_n(x,\Theta) = \sum_0^n q_i(\Theta)x^i$ *is unique, and the coefficients* $q_i(\Theta)$ *are differentiable, then*

$$\frac{\partial Q_n(\bar{\mu},\Theta)}{\partial\Theta} = 0.$$

In fact, the norm $N(\Theta)$ satisfies $\partial N/\partial\Theta = q_k(\Theta)$ and $N(\Theta) = q_0(\Theta)\mu_0 + q_1(\Theta)\mu_1 + \cdots + q_k(\Theta)\cdot\Theta + \cdots + q_n(\Theta)\mu_n$; hence

$$\sum_0^n q_i'(\Theta)\mu_i = 0.$$

COROLLARY 3. *For a segment of class* II *the norm* $N(\mu_1,\mu_2,\cdots,\mu_n)$ *is differentiable with respect to all its arguments, and the coefficients of the extremal polynomial are continuous functions of these arguments.*

This follows from (33) and the uniqueness of the extremal polynomial.

§2. Segments of class I

We now introduce the concept of a fictitious distribution of a segment. Let the segment $(\mu_i)_0^n$ have the structure $\mu_k = \sum_{i=1}^m \Delta_i\rho_i^k$ $(k = 0, 1, \cdots, n)$, where (ρ_i) and (Δ_i) are real numbers, subject to no restrictions except $\Delta_i \neq 0$ and $\rho_i \neq \rho_l$ for $l \neq i$. Attaching to each ρ_i the sign of the corresponding Δ_i, we obtain the distribution $(\overset{\pm}{\rho}_i)_1^m$. If this distribution is not identical with the faithful distribution of the segment, we call it fictitious. The polynomial $\prod_1^m (x - \rho_i)$ is called the fictitious resolvent.

According to the definition (p. 37), for the faithful representation $(\overset{\pm}{\sigma}_i)_1^s$ of a segment $(\mu_i)_0^n$ of class I we have s not greater than the degree of the "squared resolvent" (Theorem 14). We shall show that such a segment is uniquely determined by the coefficients of its resolvent.

REMARK. If the segment $(\mu_i)_0^n$ (non-absolutely-monotonic) is successively extended in the best way, $(\mu_p)_0^\infty$ has a step-function integrator $g(t)$ (Theorem 2). Then if $(\overset{\pm}{\sigma}_i)_1^s$ is the distribution of the resulting moment sequence, we have:

1) $(\mu_p)_1^\infty$ does not have $\sigma = 0$ as a node of its distribution, since

$$\mu_p = \int_0^1 t^p dg(t) = \int_0^1 t^{p-1} dh(t) = \sum_{i=1}^s \sigma_i^{p-1}(\sigma_i\delta_i) \quad (p = 1, 2, \cdots);$$

2) $(\mu_{p,1})_{p=0}^{\infty}$ does not have $\sigma = 1$ as a node of its distribution, since

$$\mu_{p,1} = \int_0^1 t^p(1-t)\,dg(t) = \int_0^1 t^p dg_1(t)$$

$$= \sum_{i=1}^{s} \sigma_i^p[(1-\sigma_i)\delta_i] \quad (p = 0, 1, 2, \cdots);$$

3) $(\mu_{p,1})_1^{\infty}$ does not have 0 or 1 as a node of its distribution.

This is clear from 1 and 2.

Now consider the finite segments $(\mu_i)_0^n$, $(\mu_i)_1^n$, $(\mu_{i,1})_0^n$, $(\mu_{i,1})_1^n$. If the first is constructed from the distribution $(\overset{\pm}{\sigma_i})_1^s$, then the second is constructed from the same distribution, possibly with $\sigma_1 = 0$ omitted; the third, possibly with $\sigma_s = 1$ omitted; the fourth, with both $\sigma_1 = 0$ and $\sigma_s = 1$ omitted (if they occurred in the first place).

This remark is similar to the remark on pp. 24-25 about absolutely monotonic segments, except for the essential difference that here $(\overset{\pm}{\sigma_i})_1^s$ is the faithful distribution only for $(\mu_i)_0^n$, whereas for $(\mu_i)_1^n$ the distribution $(\overset{\pm}{\sigma_i})_2^s$ is in general a fictitious distribution; and similarly for the other two segments.

Theorem 22 [7]. *If the segment* $(\mu_i)_0^n$ *belongs to class* I, *and if* $(\overset{\pm}{\sigma_i})_1^s$ *is its faithful distribution, then the coefficients of the resolvent*

$$R_s(x) = \prod_1^s (x - \sigma_i) = \alpha_0 + \alpha_1 x + \cdots + \alpha_{s-1}x^{s-1} + x^s$$

are determined by a system of s equations with nonvanishing determinant.

1. Suppose that all the nodes σ_i are interior to $[0,1]$.

By hypothesis, $2s \leq n$. Form the s equations in the unknowns x_i:

$$(*_1) \qquad \begin{aligned} &\mu_0 x_0 + \mu_1 x_1 + \cdots + \mu_{s-1}x_{s-1} + \mu_s = 0 \\ &\mu_1 x_0 + \mu_2 x_1 + \cdots + \mu_s x_{s-1} + \mu_{s+1} = 0 \\ &\cdots\cdots\cdots\cdots\cdots\cdots\cdots\cdots \\ &\mu_{s-1}x_0 + \mu_s x_1 + \cdots + \mu_{2s-2}x_{s-1} + \mu_{2s-1} = 0. \end{aligned}$$

Since $\mu_i = \sum_{k=1}^s \delta_k \sigma_k^i$, the system $(*_1)$ can be rewritten in the form

$$(*_1') \qquad \sum_{i=1}^{s} \delta_i \sigma_i^k L(\sigma_i) = 0, \quad (k = 0, 1, \cdots, s-1),$$

where $L(x) = x_0 + x_1 x + x_2 x^2 + \cdots + x_{s-1}x^{s-1} + x^s$; here the unknowns are $L(\sigma_i)$. The system $(*_1')$ is homogeneous and has a nonvanishing determinant; consequently

$$(*_1'') \qquad L(\sigma_i) = 0 \quad (i = 1, 2, \cdots, s).$$

The system $(*_1'')$ is inhomogeneous and has a Vandermonde determinant. Consequently the x_i are uniquely determined. But the system $(*_1)$ always has a solution (x_i are the coefficients of the resolvent); if it has a second solution (β_i), this solution must also satisfy $(*_1'')$, which is impossible.

Thus the system $(*_1)$ has a unique solution, and therefore its determinant is not zero:

$$(34) \qquad \Delta(\mu_0, \mu_{2s-2}) \neq 0.$$

Therefore in case 1 condition (34) is a necessary condition for $(\mu_i)_0^n$ to belong to class I.

2. Suppose that $\sigma_1 = 0$ and $\sigma_s < 1$; then $2s \leqq n + 1$. The resolvent has the form $x(\alpha_1 + \alpha_2 x + \cdots + \alpha_{s-1}x^{s-2} + x^{s-1})$, a faithful resolvent.

As we remarked, the segment $(\mu_i)_1^n$ has the fictitious distribution $(\overset{\pm}{\sigma}_i)_2^s$, where $(\alpha_i)_1^{s-1}$ are the coefficients of its fictitious resolvent; however, this fact does not affect our further discussion at all, since it affects only the structure of the μ_i and not the existence of an extremal polynomial of a particular type.

Form the system of $s-1$ equations in the unknowns $(\alpha_i)_1^{s-1}$:

$$(*_2) \qquad \begin{array}{c} \mu_1\alpha_1 + \mu_2\alpha_2 + \cdots + \mu_{s-1}\alpha_{s-1} + \mu_s = 0 \\ \cdot\ \cdot\ \cdot\ \cdot\ \cdot\ \cdot\ \cdot\ \cdot\ \cdot\ \cdot\ \cdot\ \cdot\ \cdot\ \cdot\ \cdot\ \cdot\ \cdot \\ \mu_{s-1}\alpha_1 + \mu_s\alpha_2 + \cdots + \mu_{2s-3}\alpha_{s-1} + \mu_{2s-2} = 0 \end{array}$$

with determinant $\Delta(\mu_1, \mu_{2s-3})$. Here we have case 1 and obtain the necessary condition

$$(35) \qquad \Delta(\mu_1, \mu_{2s-3}) \neq 0.$$

3. Suppose that $\sigma_1 > 0$ and $\sigma_s = 1$. Then $2s \leqq n + 1$. The segment $\mu_{0,1}, \mu_{1,1}, \cdots, \mu_{n-1,1}$ has the (fictitious) distribution $(\overset{\pm}{\sigma}_i)_1^{s-1}$. Let its fictitious resolvent be $\beta_0 + \beta_1 x + \cdots + \beta_{s-1}x^{s-2} + x^{s-1}$.

We have the system of $s-1$ equations

$$(*_3) \qquad \begin{array}{c} \mu_{0,1}\beta_0 + \cdots + \mu_{s-2,1}\beta_{s-2} + \mu_{s-1,1} = 0 \\ \cdot\ \cdot\ \cdot\ \cdot\ \cdot\ \cdot\ \cdot\ \cdot\ \cdot\ \cdot\ \cdot\ \cdot\ \cdot\ \cdot\ \cdot\ \cdot \\ \mu_{s-2,1}\beta_0 + \cdots + \mu_{2s-4,1}\beta_{s-2} + \mu_{2s-3,1} = 0. \end{array}$$

The same considerations as in the first case lead to the condition for

(β_i) to be unique:

(36) $$\Delta(\mu_{0,1}, \mu_{2s-4,1}) \neq 0.$$

After finding (β_i) we can find the faithful resolvent of the original segment $(\mu_i)_0^n$:

$$\sum_{i=0}^{s} \alpha_i x^i = (x-1) \sum_{i=0}^{s-1} \beta_i x^i.$$

4. Suppose $\sigma_1 = 0$, $\sigma_s = 1$. Then $2s \leq n+2$.

The faithful resolvent of $(\mu_i)_0^n$ is

$$R_s(x) = x(x-1)(\beta_0 + \beta_1 x + \cdots + \beta_{s-3}x^{s-3} + x^{s-2}) = \sum_0^s \alpha_i x^i.$$

The segment $\mu_{1,1}; \mu_{2,1}; \cdots; \mu_{n-1,1}$ has the (fictitious) distribution $(\overset{\pm}{\sigma}_i)_2^{s-1}$. To determine $(\beta_i)_0^{s-3}$ we write the system of $s-2$ equations

$$(*_4) \qquad \begin{aligned} \mu_{1,1}\beta_0 + \cdots + \mu_{s-2,1}\beta_{s-3} + \mu_{s-1,1} &= 0 \\ \cdots\cdots\cdots\cdots\cdots\cdots\cdots\cdots \\ \mu_{s-2,1}\beta_0 + \cdots + \mu_{2s-5,1}\beta_{s-2} + \mu_{2s-4,1} &= 0. \end{aligned}$$

The same considerations as before lead to the condition

(37) $$\Delta(\mu_{1,1}, \mu_{2s-5,1}) \neq 0.$$

After finding (β_i) we can find (α_i).

The present theorem is analogous to Theorem 9.

COROLLARY 1. *A segment $(\mu_i)_0^n$ of class* I *with faithful distribution $(\overset{\pm}{\sigma}_i)_1^s$ can be extended in a natural way "according to its resolvent," starting from μ_s; that is, the numbers $\mu_s, \mu_{s+1}, \cdots$ are determined from the conditions*

$$\alpha_0\mu_0 + \cdots + \alpha_{s-1}\mu_{s-1} + \mu_s = 0;$$

$$\alpha_0\mu_1 + \cdots + \alpha_s\mu_s + \mu_{s+1} = 0;$$

and so on; and the sequence $\mu_{n+1}, \mu_{n+2}, \cdots$ so obtained provides the sequence of best extensions of $(\mu_i)_0^n$.

We note that for a segment of class II the best extension is the extension "according to its resolvent."

COROLLARY 2. *If $(\mu_i)_0^n$ is a segment of class* I *with faithful distribution $(\overset{\pm}{\sigma}_i)_1^s$, then (supposing for definiteness that $0 < \sigma_i < 1$ for $i = 1, 2, \cdots, s$) all determinants $\Delta(\mu_m, \mu_{m+2s}) = 0$ $(m = 0, 1, 2, \cdots)$.*

COROLLARY 3. *If the segment* $(\mu_i)_0^n$ *satisfies the hypotheses of one of the four cases of Theorem* 22, *i.e.* (34), (35), (36) *or* (37), *and if the resolvent obtained, which may be fictitious, is such that all its zeros are in the interval* $[0,1]$, *if after solving the system of equations* $\sum_{i=1}^{s}\delta_i\sigma_i^k=\mu_k$ $(k=0,1,\cdots,s)$ *for* (δ_i) *and assigning to each* σ_i *the sign of its corresponding* δ_i *the resulting distribution* $(\overset{\pm}{\sigma}_i)_1^s$ *is of degree at most* n (*p.* 36), *then these conditions are necessary and sufficient for the given segment to be of class* I.

EXAMPLE 24. $(\mu_i)_0^n=1,2,\cdots,n+1$; find the class of the segment. We have

$$\Delta(\mu_s,\mu_{s+2})=\begin{vmatrix} s+1 & s+2\\ s+2 & s+3\end{vmatrix}=-1 \text{ for } s=0,1,\cdots,n-2,$$

and all the determinants of higher order are zero.

Thus if this segment is of class I, its resolvent $R_2(x)$ has coefficients $\alpha_0=1$; $\alpha_1=-2$, i.e. $R_2(x)=(1-x)^2$. But then the best extension of the segment is $\mu_{n+1}^*=n+2,\mu_{n+2}^*=n+3,\cdots$, which is impossible since we would have $\mu_p\to\infty$. Therefore the segment is of class II. It is easily verified, by forming the sequence of differences $(\mu_{0,i})_0^n$, that the principal polynomial is $(-1)^nT_n(x)$.

EXAMPLE 25. $(\mu_i)_0^n=1/4,3/16,\cdots,(2^{n+1}-1)/2^{2n+2}$; here $\mu_i=\frac12(\frac12)^i-\frac14(\frac14)^i$, and therefore the segment has nodal structure with distribution $(\overset{-}{\tfrac14},\overset{+}{\tfrac12})$. If the segment is of class I this is its only faithful distribution (p. 34). We now determine the degree m of this distribution. The numbers $\overset{-}{\tfrac14},\overset{+}{\tfrac12}$ are a pair of adjacent nodes of the polynomial $T_6(x)$, and consequently $m=6$ (§1, Chapter 3, and also Theorem 23).

Thus the segment is of class I for $n\geqq 6$; for $2\leqq n\leqq 5$ the segment is of class II and $\overset{-}{\tfrac14},\overset{+}{\tfrac12}$ is a fictitious distribution. For $n=1$ the segment is absolutely monotonic (amorphous), since $(\mu_i)_0^1=1/4,1/16$ has an interval of best extension.

EXAMPLE 26. $(\mu_i)_0^n=0,a,a^2,\cdots,a^n$. This segment has the nodal structure $1-1;a-0;a^2-0;\cdots;a^n-0$ with two nodes $(\overset{-}{0},\overset{+}{a})$. Consequently if the segment is of class I this distribution is its faithful distribution. This occurs if and only if $1\geqq a\geqq\sin^2(\pi/2n)$. In all other cases (for $a\neq 0$) the segment is of class II (Theorem 24).

§3. Faithful and fictitious distributions of a segment

In Chapter I we introduced the concept of a distribution, i.e. of a set of numbers $(\overset{\pm}{\sigma}_i)_1^s$ in the interval $[0,1]$, where each σ_i is assigned one of the signs $+$ or $-$. Such a distribution is associated with a polynomial $Q(x)$ for which

$$(38) \qquad Q(\overset{+}{\sigma}_i) = +1; \quad Q(\bar{\sigma}_i) = -1, \text{ with } \max_{[0,1]} |Q(x)| = 1,$$

and also with a segment-functional $(\mu_i)_0^n$ for which the distribution $(\overset{\pm}{\sigma}_i)_1^s$ is the faithful distribution.

We shall have to consider distributions $(\overset{\pm}{\sigma}_i)_1^s$ that are given directly, independently of any polynomial or segment; the numbers σ_i can be arbitrary, even complex; the signs associated with them can be arbitrarily assigned. When $0 \leqq \sigma_i \leqq 1$ for $i = 1, 2, \cdots, s$, a fundamental problem is to find the degree of a given distribution.

DEFINITION. The degree of a distribution is the smallest degree n for a polynomial $Q_{\underline{n}}(x)$ for which (38) is satisfied (the signs assigned to the σ_i are not all the same). The polynomial $Q_{\underline{n}}(x)$ is called a principal polynomial for the given distribution; polynomials of degree greater than n, satisfying (38), are also called polynomials of the distribution. (All these polynomials are reduced.)

The degree of a distribution depends not only on the number s of points σ_i (nodes), but also in an essential way on their positions on the interval $[0, 1]$. It is obvious that if the degree of a distribution is n, and we construct a segment $(\mu_i)_0^n$ with this distribution, i.e. $\mu_k = \sum_{i=1}^{s} \Delta_i \sigma_i^k$, where $\Delta_i > 0$ for $\overset{+}{\sigma}_i$ and $\delta_i < 0$ for σ_i, the given distribution is its faithful distribution, and the polynomial $Q_n(x)$ of the definition is a principal extremal polynomial for the segment.

We shall discuss some elementary properties of binodal distributions $(\overset{\pm}{\rho}_1, \overset{\mp}{\rho}_2)$ (opposite signs are assigned to the nodes).

THEOREM 23. 1. *If the distribution* $(\bar{\rho}_1, \overset{+}{\rho}_2)$ *is given, where* $0 < \rho_1 < \rho_2 < 1$ *(interior nodes), and* $Q_n(x)$ *is a principal polynomial for it, then there are polynomials of the distribution of degree* $n + 2$.

2. *If either* $\rho_1 = 0$ *or* $\rho_2 = 1$, *there are polynomials of the distribution of degree* $n + 1$.

In fact, for the first case there are two families of polynomials:

$$Q_{n+2}^{(1)}(x) = Q_n(x) + C[1 - Q_n(x)](x - \rho_1)^2$$

if $0 \leqq 1 - C(x - \rho_1)^2 \leqq 1$ for $0 \leqq x \leqq 1$;

$$Q_{n+2}^{2}(x) = Q_n(x) - C[1 + Q_n(x)](x - \rho_2)^2$$

if $0 \leqq 1 - C(x - \rho_2)^2 \leqq 1$ for $0 \leqq x \leqq 1$.

This statement can easily be verified directly, for example by

$$Q_{n+2}^{(1)}(x) = Q_n(x)\left[1 - C(x-\rho_1)^2\right] + C(x-\rho_1)^2.$$

Consequently, because $-1 + 2C(x-\rho_1)^2 \leqq Q_{n+2}^{(1)}(x) \leqq 1$, it is evident that $Q_{n+2}^{(1)}(x)$ is reduced.

For the second case we can take

$$Q_n(x) + C[1 - Q_n(x)]x \quad \text{and} \quad Q_n(x) - C[1 + Q_n(x)](1-x).$$

COROLLARY. *If $Q_n(x)$ is a polynomial of a given binodal distribution with interior nodes and if its degree cannot be increased by unity, then it cannot be decreased; thus $Q_n(x)$ is the only principal polynomial of the given distribution (Theorem 15).*

THEOREM 24. *A binodal distribution $(\bar{\tau}_1, \overset{+}{\tau}_2)$ [or $(\overset{+}{\tau}_1, \bar{\tau}_2)$] for which τ_1 and τ_2 are two successive nodes of the polynomial $T_n(x) = \cos n \arccos(2x-1)$ has $T_n(x)$ (or $-T_n(x)$) as unique principal extremal polynomial.*

Suppose that there is another $Q_n(x)$; then (Theorem 14)

$$Q_n(x) = T_n(x) + \phi(x)(x-\tau_1)^2(x-\tau_2)^2,$$

where $\phi(x)$ is of degree at most $n-4$, if $0 < \tau_1 < \tau_2 < 1$; but if $\tau_1 = 0$ or $\tau_2 = 1$, then

$$Q_n(x) = T_n(x) + \phi(x)x(x-\tau_2)^2$$

or

$$Q_n(x) = T_n(x) + \phi(x)(1-x)(x-\tau_1)^2,$$

and in this case $\phi(x)$ is of degree at most $n-3$. But by the properties of the polynomial $T_n(x)$, $\phi(x)$ must have at least $n-2$ changes of sign in $[0,1]$, which is impossible.

Thus the least possible degree for a second polynomial of this distribution is $n+2$ for interior nodes and $n+1$ for one boundary node; these cases can actually occur, by Theorem 23.

EXAMPLE 27. Let $\tau_1 < \tau_2$ be two consecutive nodes of $T_n(x)$. Suppose for definiteness that $T_n(\tau_1) = -1$ and $T_n(\tau_2) = +1$. Take numbers σ_1 and σ_2 so that $\tau_1 < \sigma_1 < \sigma_2 < \tau_2$. The degree of the distribution $(\bar{\sigma}_1, \overset{+}{\sigma}_2)$ can be arbitrarily large. We first show that it is always greater than n.

Let $Q_n(x)$ be a principal polynomial of the distribution. Form the linear function

$$\alpha x + \beta = \frac{\sigma_2 - \sigma_1}{\tau_2 - \tau_1}x + \frac{\sigma_1\tau_2 - \sigma_2\tau_1}{\tau_2 - \tau_1},$$

which maps the points (τ_1, τ_2) on (σ_1, σ_2). Since $0 < \alpha < 1$ and

$$\alpha + \beta = \frac{\sigma_2(1-\tau_1) - \sigma_1(1-\tau_2)}{\xi_2 - \xi_1},$$

i e. $0 < \alpha + \beta < 1$, the function $\alpha x + \beta$ maps $[0,1]$ on the subinterval $[\alpha, \alpha + \beta]$. Consequently $\overline{Q}_n(x) = Q_n(\alpha x + \beta)$ has the properties

$$\max_{[0,1]} |\overline{Q}_n(x)| = 1; \qquad \overline{Q}_n(\tau_1) = -1, \quad \overline{Q}_n(\tau_2) = +1.$$

Thus $\overline{Q}_n(x)$ is a principal polynomial of the distribution $(\bar{\tau}_1, \overset{+}{\tau}_2)$ and $\overline{Q}_n(x) \not\equiv T_n(x)$ (since $Q_n(\alpha x + \beta)$ has $s < n+1$ nodes); this is impossible by Theorem 24, and so the degree of $(\bar{\sigma}_1, \overset{+}{\sigma}_2)$ is not less than $n+2$.

Since σ_1 and σ_2 can be chosen arbitrarily close together, i.e. as close together as the distance between successive nodes of $T_p(x)$ (where $p > n$), the degree of the distribution can be arbitrarily large.

We now generalize the concept of a fictitious distribution, introduced in the preceding section.

Consider two sets of complex numbers $(\rho_i)_1^{s_1}$ and $(\Delta_i)_1^{s_1}$ satisfying the following conditions: $\rho_i \neq \rho_j$ for $i \neq j$, and the sums $\mu_k = \sum_{i=1}^{s_1} \Delta_i \rho_i^k$ are real for $k = 0, 1, \cdots, n$; then $(\rho_i)_1^{s_1}$ is called a distribution for $(\mu_i)_0^n$, and fictitious if it is not faithful.

In the special case when all Δ_i are real, the numbers ρ_i can, as before, be assigned the signs of the Δ_i; in the contrary case this cannot be done.

Example 28. Take two pairs of conjugate complex numbers, $z, \bar{z}$ and $\Delta, \bar{\Delta}$; then the segment $(\mu_k)_0^n = \Delta + \bar{\Delta};\ \Delta z + \bar{\Delta}\bar{z}; \cdots; \Delta t^n + \bar{\Delta}\bar{z}^n$ is real and has a fictitious binodal distribution; consequently $s \geqq n$ (Theorem 25).

§4. Some sufficient conditions for a segment to belong to class II

Theorem 25. *If the segment $(\mu_i)_0^n$ has the faithful or fictitious distribution $(\overset{\pm}{\rho}_i)_1^{s_1}$ then any other distribution for this segment has s_2 nodes, with $s_1 + s_2 > n + 1$. In particular, if a segment has a fictitious distribution with $s_1 \leqq n/2$ nodes, it is of class* II.

Suppose that $(\mu_i)_0^n$ has the two distributions $(\rho_i)_1^{s_1}$ and $(\lambda_j)_1^{s_2}$ with $s_1 + s_2 \leqq n + 1$. Then we have a homogeneous system of $n+1$ equations in $s_1 + s_2$ unknowns δ_i' and δ_j'':

$$(\mu_k =) \sum_{i=1}^{s_1} \delta_i' \rho_i^k = \sum_{j=1}^{s_2} \delta_j'' \lambda_j'' \quad (k = 0, 1, \cdots, n). \tag{39}$$

Taking the first $s_1 + s_2$ equations, we have a system with a Vandermonde determinant and $\rho_i \neq \lambda_j$; consequently all $\delta_i' = \delta_j'' = 0$, which is impossible.

If some of the points ρ_i and λ_j coincide, then after identifying them the number of unknowns in (39) is reduced by the same amount as the number of equations, and the same reasoning applies.

Theorem 25 can be generalized, but first we have to consider some determinants which we call determinants of Vandermonde type.

Let $\rho_1, \rho_2, \cdots, \rho_n$ be real or complex numbers, no two equal, and let $(0 <)\ k_1 < k_2 < \cdots < k_{n-1}$ be integers; then

$$V_n[(\rho_i)_1^n; (k_i)_1^{n-1}] = \begin{vmatrix} 1 & 1 \cdots & 1 \\ \rho_1^{k_1} & \rho_2^{k_1} \cdots & \rho_n^{k_1} \\ \cdot & \cdot\ \cdot\ \cdot\ \cdot & \cdot \\ \rho_1^{k_{n-1}} & \rho_2^{k_{n-1}} \cdots & \rho_n^{k_{n-1}} \end{vmatrix}$$

is a determinant of Vandermonde type; clearly $V_n[(\rho_i)_1^n; (i)_1^{n-1}]$ is an ordinary Vandermonde determinant $V_n(\rho_1, \rho_2, \cdots, \rho_n)$, which satisfies the recurrence formula

$$V_n(\rho_1, \cdots, \rho_{n-1}, \rho_n) = V_{n-1}(\rho_1, \rho_2, \cdots, \rho_{n-1}) \prod_{j=1}^{n-1} (\rho_n - \rho_j). \tag{40}$$

It is known that when the $(\rho_j)_1^n$ are positive $(0 < \rho_1 < \rho_2 < \cdots < \rho_n)$ we have $V_n[(\rho_j)_1^n; (k_j)_1^{n-1}] > 0$ [**23**]; but for our present purposes we need to remove this restriction on (ρ). Here we shall calculate determinants of Vandermonde type explicitly in two special cases.

Case 1. A determinant of the form

$$\begin{vmatrix} 1 & 1 & \cdots & 1 \\ \rho_1 & \rho_2 & \cdots & \rho_n \\ \cdot & \cdot & \cdot\ \cdot & \cdot \\ \rho_1^{k-1} & \rho_2^{k-1} & \cdots & \rho_n^{k-1} \\ \rho_1^{k+1} & \rho_2^{k+1} & \cdots & \rho_n^{k+1} \\ \cdot & \cdot & \cdot\ \cdot & \cdot \\ \rho_1^{n} & \rho_2^{n} & \cdots & \rho_n^{n} \end{vmatrix} = \frac{V_n}{k}(\rho_1, \rho_2, \cdots, \rho_n)$$

(a determinant with one omitted power) satisfies

$$\frac{V_n}{k}(\rho_1, \rho_2, \cdots, \rho_n) = V_n(\rho_1, \rho_2, \cdots, \rho_n)\, s_{n-k},$$

where $s_k = \sum \rho_{l_1}, \rho_{l_2}, \cdots, \rho_{l_k}$ with $l_1, l_2, \cdots, l_k \leq n$ and all different.

We may prove this as follows.

Form the determinant $V_{n+1}(\rho_1, \rho_2, \cdots, \rho_n, x) = \sum p_k x^k$. Then

$$\frac{V_n}{k}(\rho_1, \rho_2, \cdots, \rho_n)$$

is the minor of the element x^k in the first determinant, and consequently

$$\frac{V_n}{k}(\rho_1, \rho_2, \cdots, \rho_n) = (-1)^{n-k} p_k;$$

the roots of $\sum_0^n p_j x^j$ are $\rho_1, \rho_2, \cdots, \rho_n$, and thus $p_k = (-1)^{n-k} s_{n-k} p_n$, where s_{n-k} is the sum of all products of $n-k$ elements of $\rho_1, \rho_2, \cdots, \rho_n$. Since $\rho_n = V_n(\rho_1, \cdots, \rho_n)$, we have

$$\frac{V_n}{k}(\rho_1, \rho_2, \cdots, \rho_n) = V_n(\rho_1, \cdots, \rho_n) s_{n-k}. \tag{41}$$

Case 2. Expand the determinant with two omitted powers,

$$\frac{V_n}{k, p}(\rho_1, \rho_2, \cdots, \rho_n) = \begin{vmatrix} 1 & 1 & \cdots & 1 \\ \rho_1 & \rho_2 & \cdots & \rho_n \\ \cdot & \cdot & \cdot & \cdot \\ \rho_1^{k-1} & \rho_2^{k-1} & \cdots & \rho_n^{k-1} \\ \rho_1^{k+1} & \rho_2^{k+1} & \cdots & \rho_n^{k+1} \\ \rho_1^{p-1} & \rho_2^{p-1} & \cdots & \rho_n^{p-1} \\ \rho_1^{p+1} & \rho_2^{p+1} & \cdots & \rho_n^{p+1} \\ \cdot & \cdot & \cdot & \cdot \\ \rho_1^{n+1} & \rho_2^{n+1} & \cdots & \rho_n^{n+1} \end{vmatrix}.$$

We put

$$\frac{V_{n+1}}{k}(\rho_1, \rho_2, \cdots, \rho_n, x) = Q_{n+1}(x) = \sum_{j=0}^{n+1} q_j x^j.$$

Here $q_k = 0$ and the roots are $\rho_1, \rho_2, \cdots, \rho_n, x_{n+1}$.

Hence

$$s_{n+1-k}^{(n+1)} + x_{n+1} s_{n-k}^{(n+1)} = 0 \quad \text{and} \quad x_{n+1} = -\frac{s_{n+1-k}^{(n+1)}}{s_{n-k}^{(n+1)}}.$$

The determinant

$$\frac{V_n}{k, p}(\rho_1, \rho_2, \cdots, \rho_n)$$

is the minor of the element x^p in the determinant $(V_{n+1}/k)(\rho_1, \rho_2, \cdots, \rho_n, x)$.

Consequently

$$\frac{V_n}{k,p} = (-1)^{n+p-1} \cdot q_p = \frac{s_{n+1-p}}{q_{n+1}}, \quad \text{where} \quad q_{n+1} = \frac{V_n}{k}(\rho_1, \rho_2, \cdots, \rho_n).$$

Since

$$s_{n+1-p} = s_{n+1-p}^{(n+1)} - \frac{s_{n+1-k}^{(n+1)}}{s_{n-k}^{(n+1)}} \cdot s_{n-p}^{(n+1)},$$

we finally have, since $s_{n-k} = s_{n-k}^{(n+1)}$ (the value of $s_m^{(k)}$ is given on p. 47)

$$\text{(42)} \qquad \frac{V_n}{k,p} = V_n(\rho_1, \rho_2, \cdots, \rho_n)\left[s_{n-p+1} \cdot s_{n-k} - s_{n-k+1} s_{n-p}\right].$$

A determinant with more than two omitted powers can be expanded similarly.

We now introduce the concept of an incomplete segment with respect to a given segment.

Definition. An incomplete segment is a segment (μ_{k_i}), where $k_i = k_1, k_2, \cdots, k_r$ (the k_i are all different and are chosen from the numbers $0, 1, 2, \cdots, n$), i.e. in an incomplete segment $n + 1 - r$ elements are suppressed, and the remaining elements are taken in their original order in the segment $(\mu_i)_0^n$. In the special case $r = n + 1$ the segment is complete.

A distribution (faithful or fictitious) of an incomplete segment (μ_{k_i}) is a distribution $(\overset{\pm}{\rho}_i)_1^s$ yielding the expansion

$$\mu_{k_p} = \sum_{i=1}^{s} \Delta_i \rho_i^{k_p} \quad (p = 1, 2, \cdots, r)$$

(a system of r equations in the unknowns Δ_i), where $\overset{+}{\rho}_i$ correspond to positive Δ_i and $\bar{\rho}_i$ correspond to negative Δ_i, and the following conditions are satisfied: 1) $s \leqq r$ and 2) from the system we can select s equations with determinant different from zero.

Note that for any choice of s equations from among the r original ones the determinant of the system is of Vandermonde type with its components $(\rho_i)_1^s$ all different. If, for example, $0 < \rho_1 < \rho_2 < \cdots < \rho_s$, the second condition drops out.

Theorem 26 (generalization of Theorem 25). *An incomplete segment $(\mu_{k_i})_{i=1}^r$ cannot have two distributions $(\overset{\pm}{\rho}_i)_1^{s_1}$ and $(\overset{\pm}{\lambda}_j)_1^{s_2}$ if $s_1 + s_2 \leqq r$ and there is a nonzero determinant of Vandermonde type of order $s_1 + s_2$, formed from the powers of the combined system (ρ_i, λ_j), where the powers in the rows of the determinant are chosen from the numbers $k_1, k_2, \cdots, k_r$.*

The proof is precisely similar to the proof of Theorem 25, even in the case when the two systems (ρ_i) and (λ_j) have common elements which are identified for the proof.

As corollaries we give two simple sufficient conditions for a segment to be of class II; they also provide a lower bound for the number s of faithful nodes of a segment.

FIRST TEST. *If the segment* $(\mu_i)_0^n$ *has an incomplete segment* $(\mu_{k_i})_{i=1}^r$ *with a fictitious distribution having* m *nodes, and if the faithful distribution of the complete segment has* s *nodes, then* $s+m>r$.

In particular, if $r=n+1$ then $s+m>n+1$, and therefore if $m \leqq \frac{1}{2}n$ we have $s>\frac{1}{2}n+1$, and $(\mu_i)_0^n$ is a segment of class II.

SECOND TEST. *Suppose that the segment* $(\mu_i)_0^n$ *has zero elements* $\mu_{p_1}=0$, $\mu_{p_2}=0,\cdots,\mu_{p_m}=0$, *and that the other* $\mu_i \neq 0$. *Then the number* s *of faithful nodes satisfies* $s>m$.

Thus if $m \geqq \frac{1}{2}n+1$ the segment is of class II.

EXAMPLE 29. $(\mu_i)_0^n = 0_0,\cdots,0_{k-1},1,0_{k+1},\cdots,0_{n-1},\Theta_n$. The number s of faithful nodes satisfies $s \geqq n$ $(1 \leqq k \leqq n-1)$.

EXAMPLE 30. $(\mu_i)_0^n = 1,\cos\phi,\cos 2\phi,\cdots,\cos n\phi$. This segment has (for $\phi \neq 0$) the fictitious nodal distribution

$$\cos k\phi = \tfrac{1}{2}(e^{\phi i})^k + \tfrac{1}{2}(e^{-\phi i})^k \quad (k=0,1,\cdots,n)$$

with two fictitious (complex) nodes $\rho_1 = e^{\phi i}$ and $\rho_2 = e^{-\phi i}$. Hence $s \geqq n$.

EXAMPLE 31. $(\mu_i)_0^n = 0,\sin\phi,\sin 2\phi,\cdots,\sin n\phi$ $(\phi \neq 0)$. This segment has the binodal distribution

$$\sin k\phi = \frac{1}{2i}(e^{\phi i})^k - \frac{1}{2i}(e^{-\phi i})^k \quad (k=0,1,\cdots,n)$$

with the fictitious nodes $\rho_1 = e^{\phi i}$ and $\rho_2 = e^{-\phi i}$. Hence $s \geqq n$.

EXAMPLE 32. $(\mu_i)_0^n = 1,\rho,\cdots,\rho^{k-1},\mu_k,\rho^{k+1},\cdots,\rho^n$, where $\rho>1$ and $\mu_k \neq \rho^k$. We have $s \geqq n$ by the first test.

§5. Variable segment-functionals

THEOREM 27 (THEOREM ON CONTINUOUS DEFORMATION). *If the segment* $(\mu_i)_0^n$ *contains a variable element, say* $\mu_p = \Theta$ $(p>0)$, *whose domain is a finite interval* $\alpha \leqq \Theta \leqq \beta$ *in which the segment remains of class* II, *then the principal polynomial* $Q_n(x,\Theta) = \sum_{i=0}^n q_i(\Theta)x^i$ *is unique at each point* Θ *such that* $q_i(\Theta)$ *is continuous in* $[\alpha,\beta]$ $(i=0,1,\cdots,n)$.

We take a point $\Theta_0 \in [\alpha, \beta]$ and show that $\lim_{\Theta \to \Theta_0} q_i(\Theta)$ exists and equals $q_i(\Theta_0)$.

We recall that $Q_n(x, \Theta)$ is a reduced polynomial and consequently $|q_i| \leqq |t_i|$, where $\sum_{k=0}^{n} t_k x^k = T_n(x)$. Thus for any Θ_0 the function $q_i(\Theta)$ attains its upper and lower bounds on $[\alpha, \beta]$.

1. Suppose that each coefficient of $q_i(\Theta)$ has a limit as $\Theta \to \Theta_0$ and let $q_i(\Theta) \to p_i$ $(i = 0, 1, \cdots, n)$.

Then we also have $\lim Q_n(x, \Theta) = \sum_0^n p_i x^i = P_n(x)$; $-\infty < x < +\infty$. On an arbitrary finite interval $A \leqq x \leqq B$ we have

$$|P_n(x) - Q_n(x, \Theta)| = \left| \sum_0^n [p_i - q_i(\Theta)] x^i \right| \leqq \sum_0^n |p_i - q_i(\Theta)| \cdot M,$$

where $M = \max_{[A,B]} |x^i|$ for $i = 0, 1, \cdots, n$. However, for $|\Theta - \Theta_0| < \epsilon$ we may suppose $|p_i - q_i(\Theta)| < \epsilon/(n+1)M$ and then we have

$$|P_n(x) - Q_n(x, \Theta)| < \epsilon \text{ on } [A, B].$$

In addition, $\max_{[0,1]} |Q_n(x, \Theta)| = 1$. If there is a point x_0 on $[0, 1]$ at which $|P_n(x_0)| > 1$, then we must also have $|Q_n(x_0, \Theta)| > 1$ for sufficiently small $|\Theta - \Theta_0|$. We now show that there are points on $[0, 1]$ at which $P_n(x) = 1$. In fact, we have

$$P_n(\bar{\mu}) = \lim_{\Theta \to \Theta_0} Q_n(\bar{\mu}, \Theta) = \lim_{\Theta \to \Theta_0} N(\Theta) = N(\Theta_0)$$

since the norm is continuous (Theorem 5); but for every polynomial $R_n(x)$ we have

$$|R_n(\bar{\mu})| \leqq N(\Theta) \max_{[0,1]} |R_n(x)|,$$

and since it has already been shown that $|P_n(x)| \leqq 1$ for $0 \leqq x \leqq 1$, we must have $\max_{[0,1]} |P_n(x)| = 1$, and $P_n(x)$ is an extremal polynomial for the given segment when $\Theta = \Theta_0$. Since the extremal polynomial is unique, $P_n(x) \equiv Q_n(x, \Theta_0)$.

2. Let some $q_i(\Theta)$ not have a limit as $\Theta \to \Theta_0$. We choose a subsequence $\Theta_{k_1}, \Theta_{k_2}, \cdots, \Theta_{k_n}, \cdots$, the same for all $(q_i)_0^n$, tending to Θ_0 and such that all $q_i(\Theta)$ have limits. Let $q_i(\Theta_{k_m}) \to p_i$ as $m \to \infty$. Then $Q_n(x, \Theta_{k_m}) \to P_n(x)$, where the same argument as in part 1 shows that $P_n(x) \equiv Q_n(x, \Theta_0)$. But by taking a different subsequence $\Theta_{k_p} \to \Theta_0$ we obtain in the limit a different extremal polynomial, which is impossible.

Thus we have $\lim_{\Theta \to \Theta_0} q_i(\Theta) = q_i(\Theta_0)$ for all coefficients, and $q_i(\Theta)$ are continuous functions on $[\alpha, \beta]$. (Outside the critical interval $q_i = \pm t_i$, i.e. the continuity is evident.)

Remark. The theorem is easily generalized to the case when the segment $(\mu_i)_0^n$ contains two or more variable elements under the assumption that the k-dimensional domain of the variables is bounded and such that the segment belongs to class II for each point of this domain.

In order to make some later results more precise, we shall call a segment $(\mu_i)_0^n$ "strictly of class I" if the number of its faithful nodes is $s \leqq \frac{1}{2}n - 1$; if the number of nodes satisfies $\frac{1}{2}n - 1 < s \leqq \frac{1}{2}n + 1$, we say that the segment is of intermediate class.

Theorem 28. *If the segment $(\mu_i)_0^n$ is strictly of class* I, *and one of its elements, say μ_p $(p > 0)$ is replaced by $\mu_p + h$, the segment becomes of class* II *when $h \neq 0$.*

Let $(\overset{\pm}{\sigma}_i)_1^s$ be the faithful distribution of the original segment $(\mu_i)_0^n$, where $s \leqq \frac{1}{2}n - 1$.

Let $(\nu_i)_0^n$ be the segment obtained from the given one by replacing μ_p by $\mu_p + h$.

Let $(\overset{\pm}{\sigma}{}_i')_1^{s_1}$ be the faithful distribution of this segment.

If we suppose that $s_1 \leqq \frac{1}{2}n + 1$, the system of n equations

$$(\mu_k =) \sum_{i=1}^{s} \delta_i \sigma_i^k = \sum_{j=1}^{s_1} \delta_j' \sigma_j'^k \quad (k = 0, 1, \cdots, p-1, p+1, \cdots, n)$$

is homogeneous, and there are $s_1 + s_2 \leqq n$ unknowns δ_i and δ_i'.

We select $s_1 + s_2$ equations from this system. The determinant of the new system is either an ordinary Vandermonde determinant, or one of the same type with an omitted power; in either case, since the nodes are positive, the determinant is positive, and all the unknowns are zero.

Thus $s > \frac{1}{2}n + 1$.

Corollary 1. *A segment $(\mu_i)_0^n$ (not absolutely monotonic) with one variable element $\mu_p = \Theta$ $(p > 0)$ can have at most one point $\Theta = \Theta_p^{(0)}$ for which the segment is strictly of class* I *(when the other elements are fixed); for all other points the segment is of class* II, *and consequently the principal extremal polynomial $Q_n(x, \Theta)$ is unique for all $\Theta \neq \Theta_p^{(0)}$. Clearly $\Theta_p^{(0)}$ cannot fall outside the critical interval.*

If the (principal) extremal polynomial is not unique at the point $\Theta_p^{(0)}$ and if $\Theta_p^{(0)} \neq \Theta^*$, where Θ^* is the focus of the element μ_p, we say that $\Theta_p^{(0)}$ is a singular point for this element.

COROLLARY 2. *If the segment* $(\mu_i)_0^n$ *has a fictitious distribution with* $m \leqq \frac{1}{2}n - 1$ *nodes, there are no singular points for any* $\mu_i = \Theta$ $(i = 1, 2, \cdots, n)$.

Indeed, in this case there is an incomplete segment with n elements, and then, by the first test, $s + m > n$, i.e. $s > \frac{1}{2}n + 1$, and the segment belongs to class II.

COROLLARY 3. *If* $\mu_p = \Theta_p^{(0)}$ $(n > p > 0)$ *in the segment* $(\mu_i)_0^n$, *then for any other element* $\mu_k = \Theta_k^{(0)}$ $(k > 0)$ *or* $\mu_k = \mu_k^*$ *when the other elements are fixed (duality between singular points and foci).*

This follows immediately, since otherwise the uniqueness of the extremal polynomial would be violated.

THEOREM 29. *If the segment* $(\mu_i)_0^n$ *is strictly of class* I *and its principal polynomial* $Q_n(x) = \sum_0^n q_i x^i$ *is not unique, then both (unique!) polynomials with leading coefficients* $q_{n(\max)}$ *and* $q_{n(\min)}$ *(Theorem 21) belong to class* II.

In fact, consider the segment of class II, $\mu_0, \cdots, \mu_{n-1}, \mu_n + h$, with $h > 0$, and let its unique extremal polynomial be $Q_n(x, h)$; then $\lim_{h \to 0+} Q_n(x, h) = Q_{n(\max)}(x)$ and this polynomial is also of class II, by the theorem on continuous deformation. Similarly, for $h \to 0-$ we find that $Q_{n(\min)}(x)$ is also of class II.

EXAMPLE. Let ω_1 and ω_2 be two adjacent nodes of

$$T_{n-2}(x) = \cos(n-2)\arccos(2x-1);$$

they generate the binodal distribution $(\bar{\omega}_1, \overset{+}{\omega}_2)$ (the nodes are interior and $\frac{1}{2} \leqq \omega_1 < \omega_2$).

Then $Q_n(x) = T_{n-2}(x) - C[1 + T_{n-2}(x)](x - \omega_2)^2$ is an nth degree polynomial of this distribution (Theorem 23). Since $|Q_n(1)| < 1$, we can replace x by Ax, with $A > 1$ $(0 < C < \omega_2^{-2})$, so that

$$\max_{[0,1]} |Q_n(Ax)| = 1.$$

Then the transformed polynomial

$$Q_n(Ax) = T_{n-2}(Ax) - C[1 + T_{n-2}(Ax)](Ax - \omega_2)^2$$

is a principal polynomial of the distribution $(\bar{\omega}_1/A, \overset{+}{\omega}_2/A)$ with a fixed A and arbitrary $C > 0$ belonging to a certain interval.

In fact, the degree of this polynomial cannot be anything less for the given distribution, since in the contrary case some $L_{n-1}(x)$ would have a distribution containing the nodes $\bar{\omega}_1/A$ and $\overset{+}{\omega}_2/A$, and consequently $L_{n-1}(x/A)$ would have the nodes $\bar{\omega}_1, \overset{+}{\omega}_2$, which is impossible by Theorem 24.

Therefore the principal polynomial is not unique in this case.

Now if we construct any segment $(\mu_i)_0^n$ with the distribution $(\bar{\omega}_1/A, \overset{+}{\omega}_2/A)$, we obtain a segment that is strictly of class I and precisely of degree n, and its principal polynomial is not unique. Then μ_n is a singular point of the nth element with respect to the other elements.

CHAPTER III

FAMILIES OF EXTREMAL POLYNOMIALS

§1. Construction of segments with given distribution

Consider a distribution $(\overset{\pm}{\sigma_i})_1^s$ of degree n, belonging to class II, i.e. $0 \leqq \sigma_1 < \sigma_2 < \cdots < \sigma_{s-1} < \sigma_s \leqq 1$, where $\frac{1}{2}n + 2 \leqq s \leqq n + 1$.

If the signs attached to two consecutive points σ_i, σ_{i+1} are the same, we say that there is a *permanence*, and call the interval (σ_i, σ_{i+1}) an *interval of permanence*; we denote the number of permanences in the distribution by p.

If the signs attached to two consecutive points are different, we say that there is an *alternation*, and call the corresponding interval an *interval of alternation*; we denote the number of alternations by q, and suppose that $q > 0$.

Clearly

$$p + q = s + 1. \tag{43}$$

We call the numbers n, s, p the *characteristics of the distribution*, and we call this set of numbers, denoted by $[n, s, p]$, its *passport*. These concepts, and the corresponding terms, will also be applied to the polynomials and segments corresponding to the distribution $(\overset{\pm}{\sigma_i})_1^s$.

Let $Q_n(x)$ be a reduced polynomial and let $(\overset{\pm}{\sigma_i})_1^s$ be its (complete) distribution. The passport of the polynomial means the passport $[n, s, p]$ of its distribution. We have the evident inequality

$$s + p \leqq n + 1 \tag{44}$$

(which follows by counting the zeros of $Q_n'(x)$).

In particular, when $s = n + 1$ we have $p = 0$; when $s = n$ we may have either $p = 0$ or $p = 1$; etc.

The passport of the polynomials $\pm T_n(x)$ is $[n, n+1, 0]$.

Using the distribution of the polynomial $Q_n(x)$ we construct the segment $(\mu_i)_0^n$ by taking $\mu_k = \sum_{i=1}^s \delta_i \sigma_k^i$ $(k = 0, 1, \cdots, n)$ and $\delta_i > 0$ (< 0) if σ_i carries a $+$ $(-)$ sign; the absolute values of the numbers δ_i can be taken arbitrarily (but not zero). For all such segments $Q_n(x)$ is the (unique) principal extremal polynomial; $(\overset{\pm}{\sigma_i})_1^s$ is the faithful distribution of the segment, and its passport $[n, s, p]$ is also called the *passport of the segment.*

If a segment $(\mu_k)_0^n$ is such that its passport (that is, the passport of its faithful representation) $[n, s, p]$ does not coincide with the passport of its principal polynomial $Q_n(x)$, it must be the case that the number s_1 of nodes of $Q_n(x)$ is greater than s, and then, generally speaking, the number of permanences in the passport of the polynomial will be decreased. Let this passport be $[n, s_1, p_1]$.

We emphasize that we consider passports only for segments (and polynomials) of class II.

Suppose that the segment $(\mu_i)_0^n$ has a fictitious distribution (p. 60) with m fictitious nodes. We showed (p. 64) that $s + m > n + 1$. Then by (44)

$$p \leqq m - 1. \tag{45}$$

Let us find the characteristics and passports of some simple segments of class II with one variable element.

EXAMPLE 33. $(\mu_i)_0^n = 1, \rho, \rho^2, \cdots, \rho^{n-1}, \rho^n + \Theta$, where $\rho > 1$ and Θ is variable. By Theorem 26, we have $s \geqq n$, and consequently $s = n$ in the critical interval (Example 20). In addition, since when the segment is decomposed according to any n nodes $(\sigma_i)_1^n$, where $0 \leqq \sigma_i \leqq 1$, the weights δ_i of the σ_i have alternating signs, we certainly have $p = 0$. Thus the passport of this segment is constant in the critical interval and is $[n, n, 0]$.

EXAMPLE 34. $(\mu_i)_0^n = 0_0, \cdots, 0_{n-2}, 1_{n-1}, \Theta$, where Θ is arbitrary. We have already shown (Example 29) that when Θ is in the critical interval the segment has characteristic $s = n$. When the segment is decomposed in terms of any $(\sigma_i)_1^n$, where $0 \leqq \sigma_i \leqq 1$ and $\sigma_i \neq \sigma_j$, we obtain alternating signs for the δ_i; hence the passport of this segment is $[n, n, 0]$ (by continuous deformation).

EXAMPLE 35. $(\mu_i)_0^n = 1, 1/2, 1/2^2, \cdots, 1/2^{n-1}, 1/2^n + \Theta$, where Θ is arbitrary and n is even. By Theorem 26, when $0 < \Theta < \Theta''$ there are $s = n$ faithful nodes. When the segment is decomposed in terms of the faithful nodes $(\sigma_i)_1^n$ we necessarily have $\sigma_{n/2+1} < 1/2 \leqq \sigma_{n/2+2}$, and then there is always one permanence, namely for the nodes $\sigma_{n/2+1}, \sigma_{n/2+2}$. Therefore $p = 1$ and the passport of the segment is $[n, n, 1]$.

The preceding example shows that the passport does not completely characterize the distribution, the polynomial or the segment, because of the insufficiency of the characteristic p, which gives only the number of permanences, but not their location in the system (σ_i). In some problems the passport is nevertheless sufficient; but in others we require a detailed investigation of the characteristic p. In the latter case we shall write $P = p(k_1, k_2, \cdots, k_p)$ instead of p, where the integers k_i are the indices of the intervals of permanence among the set of all intervals between nodes $[(\sigma_i, \sigma_{i+1})$ is the ith interval$]$. For example, in the passport $[n, s, 2(1, s-1)]$ the characteristic $2(1, s-1)$ means that the distribution has $p = 2$ permanences, and the intervals of permanence are the first and $(s-1)$th.

THEOREM 30. *For every polynomial with s nodes in the set* $\{Q_n(x)\}$ *of polynomials of class* II *there is a segment* $(\mu_i)_0^n$ *of fictitious nodal structure with the minimum number of fictitious nodes* $(\rho_i)_1^m$ *such that the given* $Q_n(x)$ *is the principal polynomial of the segment, and the segment and the polynomial have the same passport.*

Let $[n, s, P]$ be the passport of the polynomial $Q_n(x)$ and let $(\overset{\pm}{\sigma}_i)_1^s$ be its distribution. For the proof it is enough to show how to choose the fictitious nodes (ρ_i) of the segment in relation to the faithful nodes (σ_i) of the polynomial.

We shall present a somewhat more convenient method for this choice. Recall that $m + s > n + 1$ (p. 70), and consequently $m_{\min} = n + 2 - s$. We have $0 \leqq \sigma_1 < \sigma_2 < \cdots < \sigma_{s-1} < \sigma_s \leqq 1$.

1. In each of the p intervals of permanence we place one ρ, and we select the other $n + 2 - (s + p) = m - p$ (> 0) ρ's arbitrarily, but greater than unity. We combine the set $(\rho_i)_1^m$ with $(\sigma_i)_1^s$, renumber these points in increasing order, and denote them by $(\lambda_i)_1^{n+2}$; here we always have $\lambda_{n+2} = \rho_m > 1$. We decompose the segment $1, \lambda_{n+2}, \lambda_{n+2}^2, \cdots, \lambda_{n+2}^n$ in terms of the nodes $(\lambda_i)_1^{n+1}$, i.e. we solve the system of $n+1$ equations

$$\sum_{i=1}^{n+1} x_i \lambda_i^k = \lambda_{n+2}^k \quad (k = 0, 1, \cdots, n). \tag{46}$$

The system can be solved with alternating signs, i.e. $x_{n+1} > 0$, $x_n < 0$, etc. After solving, we separate the σ's and ρ's and transpose to the right side of (46) all terms containing ρ's and their zeroth powers, and denote the right-hand sides by μ_k $(k = 0, 1, \cdots, n)$. Then one of the segments $(\mu_i)_0^n$ and $(-\mu_i)_0^n$ (and only one) is the required segment. In fact, $(\mu_i)_0^n$ has the fictitious distribution $(\overset{\pm}{\rho}_i)_1^m$, where each ρ_i has the sign of its corresponding $-x_i$ in (46). The faithful distribution $(\overset{\pm}{\sigma}_i)_1^s$ of the segment is the same as the distribution of $Q_n(x)$, since after removing the nodes ρ both adjacent numbers σ, and only these, yield intervals of permanence, as required; it remains to choose the sign of the segment $\pm$ $(\mu_i)_0^n$ so that the signs attached to the σ's in solving (46) coincide with the signs of $(\overset{\pm}{\sigma}_i)$ in the distribution of $Q_n(x)$.

2. We distribute the numbers $(\rho_i)_1^m$ one at a time in the intervals of permanence of $(\overset{\pm}{\sigma}_i)_1^s$, and the remaining (ρ) two at a time in any intervals of alternation. If $m-p$ is even $(m>p)$, all (ρ_i) lie in $[0,1]$; if $m-p$ is odd, we take $\rho_m>1$. The choice of the segment $(\mu_i)_0^n$ and the proof of its correctness proceed just as in case 1.

3. In each interval of permanence of $(\overset{\pm}{\sigma}_i)_1^s$ we put an odd number of nodes ρ_i, and in each interval of alternation we put an even number or none. The proof is similar.

Thus every $Q_n(x)$ is the principal polynomial of a segment with the same passport $[n,s,p]$ with a fictitious distribution containing $m=n+2-s$ fictitious nodes.

We now indicate still another method of constructing a very simple segment for a given $Q_n(x)$; it is appropriate and convenient in the case $p=0$.

THEOREM 31. *If the passport of a given $Q_n(x)$ is $[n,s,0]$, a segment for which $Q_n(x)$ is the principal polynomial can be constructed as follows: $\mu_k=0$ for $k<s-1$; $\mu_{s-1}=\pm 1$; for $k\geqq s$ the remaining μ_k are determined uniquely.*

In fact, let $(\overset{\pm}{\sigma}_i)_1^s$ be the distribution of $Q_n(x)$ with the signs of the σ_i alternating throughout. Consider the system of s equations in the s unknowns δ_i:

$$\sum_1^s \delta_i=0;\quad \sum_1^s \delta_i\sigma_i=0;\ \cdots;\quad \sum_1^s \delta_i\sigma_i^{s-2}=0;\quad \sum_1^s \delta_i\sigma_i^{s-1}=1.$$

This system can be solved with alternating signs ($\delta_s>0$; $\delta_{s-1}<0$; etc.).

The remaining numbers μ can be determined from the δ_i that have been found as follows:

$$\mu_s=\sum_1^s \delta_i\sigma_i^s,\ \cdots,\mu_n=\sum_1^s \delta_i\sigma_i^n.$$

The required segment is $(+\mu_i)_0^n$ or $(-\mu_i)_0^n$ according as the number σ_s in the distribution of $Q_n(x)$ has a $+$ or $-$ sign.

§2. Parametrization of families of polynomials

THEOREM 32. *If an irreducible segment $(\mu_i)_0^n$ with $\mu_p=\Theta$ $(p>0)$ remains of class II near the point $\Theta=\Theta_0$ and if at this point the segment has s faithful nodes $(\sigma_i)_1^s$ which are also the complete set of nodes of the principal polynomial $Q_n(x,\Theta_0)$, then near Θ_0 $(\Theta=\Theta_0+\Delta\Theta)$ the number of nodes of the segment, and of the polynomial, cannot change.*

Note that if we write $P_n(x,\alpha) = \sum_0^n p_i(\alpha)x^i$, and the functions $p_i(\alpha)$ are continuous in an α-interval, then in any finite interval $A \leq x \leq B$ the polynomial $\sum_0^n p_i(\alpha + \Delta\alpha)x^i$ satisfies $|P_n(x,\alpha + \Delta\alpha) - P_n(x,\alpha)| < \epsilon$ when $|\Delta\alpha| < \vartheta$, and then the roots $\sigma_i(\alpha)$ of $P_n(x,\alpha)$ are continuous functions of α; and if these roots are real and simple, then $P_n(x,\alpha)$ and $P_n(x,\alpha + \Delta\alpha)$ have the same number of real roots.

According to what we proved in Chapter I, the polynomial

$$\frac{\partial Q_n(x,\Theta)}{\partial x} x(x-1) = P_{n+1}(x,\Theta)$$

has continuous coefficients $p_i(\Theta)$, and consequently has its roots continuous (with respect to Θ). But $P_{n+1}(x,\Theta)$ is a multiple of the resolvent, and consequently $\sigma_i(\Theta)$ are roots of the resolvent, hence continuous functions; and since they are real and simple, the number of them on $[0,1]$ is constant. (If $x = 0$ and $x = 1$ are roots of the resolvent, they must be fixed.)

Corollary 1. *Since the nodes of the polynomial $Q_n(x,\Theta)$ (these are also nodes of the segment) undergo small variations under small changes in Θ, the weights δ_i in the structural system of equations also vary continuously.*

Corollary 2. *If a given segment $(\mu_i)_0^n$ with fixed elements has passport $[n,s,P]$ coinciding with the passport of its principal polynomial, then the original characteristic P is also preserved under sufficiently small variations of the elements of the segment that preserve the number s of its nodes.*

In fact, the passport of a segment can change only when at least one weight becomes zero, i.e. when the number of nodes changes, a possibility that we have excluded.

Corollary 3. *If the initial segment $(\mu_i)_0^n$ has a fictitious distribution with $n+2-s$ fictitious nodes, arranged as in version 1 of Theorem 30, then the same fictitious nodes can be retained for sufficiently small variations of the elements $(\mu_i)_0^n$, with only their weights Δ_j varying, and indeed continuously.*

We have a segment $(\mu_i)_0^n$ with fixed elements, which we call the initial segment. Let m of its elements $\mu_{k_0}, \mu_{k_1}, \cdots, \mu_{k_{m-1}}$ (where $n/2 + 1 < m < n+1$) be chosen so that $\mu_{k_0} = \mu_0$, but otherwise arbitrarily ($k_0 < k_1 < \cdots < k_{m-1}$). These elements are fixed and form a *basis* for the variable segment that is obtained if the remaining $l = m + 1 - m$ elements are replaced by variables $\Theta_1, \Theta_2, \cdots, \Theta_l$, numbered in the order of their original indices. The initial values of the Θ_i in the segment $(\mu_i)_0^n$ will be denoted by $\theta_1^{(0)}, \theta_2^{(0)}, \cdots, \theta_l^{(0)}$. We denote such a variable segment by $[(\mu_{k_i})_{i=0}^{m-1}; (\theta_i)_1^l]$, and then $[(\mu_{k_i})_0^{m-1}; (\theta_i^{(0)})_1^l]$ coincides with $(\mu_i)_0^n$. Each set of values of the θ_i defines a point of an l-dimensional space $M(\theta_1, \cdots, \theta_l)$; we call the point $M_0(\theta_1^{(0)}, \cdots, \theta_l^{(0)})$ the initial point. Each point of this space defines a unique segment, which is a particular value of the variable segment.

Let the initial segment $(\mu_i)_0^n$ be such that its passport $[n, s, P]$ coincides with the passport of its principal polynomial $Q_n(x)$, the initial polynomial. We choose a basis for the segment as described above; the resulting variable segment $[(\mu_{k_i})_0^{m-1}; (\theta_i)_1^l]$ has $l = n + 1 - m$ variable elements.

By the theorem on continuous deformation of a polynomial, and its corollary, every point in a sufficiently small neighborhood of the initial point M_0 corresponds to a single segment and a single polynomial (of class II), the principal polynomial of the segment and with the same passport $[n, s, P]$. Any $(n + 1 - m)$-dimensional region of such points, in which both the segment and its principal polynomial have the same passport, is called a *region of invariant passport* (of the polynomial and segment) and denoted by $M_l[n, s, P]$, or for short by M_l.

It is clear that M_0 is an interior point of this region. When the boundary of M separates it from the region in which the number of nodes of the polynomial and segment is s_1, then when $s_1 > s$ the boundary points do not belong to M_l for the polynomial, but do belong to it for the segment; if $s_1 < s$, the boundary points belong to M_l for the polynomial, but not for the segment.

We shall answer the following question: what conditions ensure that the region of invariant passport is such that different (interior) points of the region, and consequently different segments, correspond to *different* polynomials, which are the principal polynomials of the segments?

If (and only if) this requirement of one-to-one-ness is satisfied, we say that the l variable elements θ_i are *independent* elements of the segment.

THEOREM 33. *Let the passport $[n, s, P]$ of the initial segment $(\mu_i)_0^n$ coincide with the passport of its principal polynomial $Q_n(x)$; then if $(\mu_{k_i})_{i=0}^{m-1}$ is a basis, with $k_0 = 0$, the variables $(\theta_i)_1^l$ are independent if $m = s$, but dependent if $m < s$.*

We first show that when $m = s$ the initial polynomial $Q_n(x)$ cannot be continued into the region of invariant passport beyond the initial point M_0. In fact, let $(\overset{\pm}{\sigma}_i)_1^s$ be the distribution of $Q_n(x)$ and let $(\mu_{k_i})_{i=1}^{s-1}$ be the selected basis. Then we have a system of s equations $\sum_{i=1}^s \delta_i \sigma_i^{k_p}$

$= \mu_{k_p}$ $(p = 0, 1, \cdots, s-1)$ uniquely determining the weights $(\delta_i)_1^s$, and consequently the values of all the variables are completely determined by the given nodes σ_i. Therefore $Q_n(x)$ can be extremal only at the point M_0. But since any interior point of $\{M\}_l$ can be taken as an initial point, the first conclusion of the theorem is established.

Now let $m < s$. We show that, in a neighborhood of the initial point M_0 in $(n+1-m)$-dimensional space, $Q_n(x)$ can be extended to be extremal. In fact, a necessary and sufficient condition for this is that the distribution $(\overset{\pm}{\sigma}_i)_1^s$ of the initial segment remains the same for points other than the initial point. Let $(\mu_{k_p})_{p=0}^{m-1}$ be the selected basis; then $\mu_{k_p} = \sum_{i=1}^s \delta_i \sigma_i^{k_p}$. We have to find $(\epsilon_i)_1^s$ satisfying the system $\sum_1^s (\delta_i + \epsilon_i) \sigma_i^{k_p} = \mu_{k_p}$ $(p = 0, 1, \cdots, m-1)$ or, equivalently, the system $\sum_{i=1}^s \epsilon_i \sigma_i^{k_p} = 0$ $(p = 0, 1, \cdots, m-1)$; this is an underdetermined homogeneous system $(m < s)$, and its infinite set of solutions contains one with modulus so small that the necessary conditions

$$\operatorname{Sgn}(\delta_i + \epsilon_i) = \operatorname{Sgn} \delta_i \quad (i = 1, 2, \cdots, s)$$

are satisfied. This establishes the second conclusion of the theorem.

COROLLARY 1. *If $m > s$, a basis $(\mu_{k_p})_{p=0}^{m-1}$ can be obtained from the basis $(\mu_{k_p})_{p=0}^{s-1}$ by omitting some additional elements from the initial segment $(\mu_i)_0^n$; then the remaining Θ_i, of which there are $l = n+1-m$, are clearly independent, since the $(n+1-m)$-dimensional domain of these variables is a subset of the $(n+1-s)$-dimensional domain of the first case of the theorem.*

Therefore when $m = s$ the number $l = n+1-s$ of *independent* elements of the segment is as large as possible.

COROLLARY 2. *If there is a polynomial $Q_n(x)$ with given passport $[n, s, P]$, there is a family of polynomials with this passport, containing $(n+1-s)$ variable elements $\Theta_1, \Theta_2, \cdots, \Theta_{n+1-s}$. Their elements are mutually independent.*

The latter statement means that there is an $(n+1-s)$-dimensional domain in the space of these variables, such that two distinct points correspond to different polynomials of the family. In fact, if we construct the segment $(\mu_i)_0^n$ with principal polynomial $Q_n(x)$ and the same passport; and then, selecting a basis $(\mu_{k_p})_{p=0}^{s-1}$, go over to the variable segment $[(\mu_{k_p})_0^{s-1}, (\Theta_i)_1^{n+1-s}]$, then $(\Theta_i)_1^{n+1-s}$ is a system of independent elements, and consequently in an $(n+1-s)$-dimensional neighborhood of the initial point $(\Theta_i^0)_1^{n+1-s}$ the variable segment defines the family of its principal polynomials, which depends on the same elements. We denote it by $Q_n(x, \Theta_1, \Theta_2, \cdots, \Theta_{n+1-s})$; then $Q_n(x, \Theta_1^{(0)}, \Theta_2^{(0)}, \cdots, \Theta_{n+1-s}^{(0)}) = Q_n(x)$. These elements, which are independent for the segment, are also independent for the polynomial.

We shall discuss later the question of whether the family

$$Q_n(x, \Theta_1, \Theta_2, \cdots, \Theta_{n+1-s})$$

contains all polynomials of the given passport.

The method just explained for constructing a variable segment is, as will be clear from what follows, one of the most useful methods for investigating polynomials of given passport; but there are other methods not covered by it. We now discuss one of them.

Let $Q_n(x)$ be a polynomial of passport $[n, s, P]$. Using its distribution $(\overset{\pm}{\sigma_i})_1^s$ we construct a segment $(\mu_i)_0^n$ with fictitious distribution $(\overset{\pm}{\rho_j})_0^{m-1}$ $(m = n + 2 - s)$, where the numbers $(\rho_i)_0^{m-1}$ are chosen according to version 1 of Theorem 30, and then fixed. Then $\mu_k = \sum_{j=0}^{m-1} \Delta_j^{(0)} \rho_j^k$ $(k = 0, 1, \cdots, n)$, where we may assume $\Delta_0^{(0)} = 1$ (by termwise division), and then we replace all other $\Delta_j^{(0)}$ by variables Δ_j. We denote the segment by $[\mu_i(\Delta_1, \Delta_2, \cdots, \Delta_{n+1-s})]_{i=0}^n$. The point $\mu_0(\Delta_1^{(0)}, \Delta_2^{(0)}, \cdots, \Delta_{n+1-s}^{(0)})$ in the $(n+1-s)$-dimensional space of the variables Δ_j is called the initial point. The region in this space whose interior points correspond to segments and polynomials of passport $[n, s, P]$ is called the region of invariant passport and denoted by $\{M\}_{n+1-s}$. Such a region necessarily exists by the theorem on continuous deformation.

We call the Δ_j independent elements of the segment

$$[\mu_i(\Delta_1, \Delta_2, \cdots, \Delta_{n+1-s})]_{i=0}^n$$

if and only if for every two points the corresponding segments have *different* principal polynomials.

THEOREM 34. *Suppose that the initial segment $(\mu_i)_0^n$ with fictitious distribution $(\rho_i)_0^{n+1-s}$ and its principal polynomial $Q_n(x)$ have the same passport $[n, s, P]$; in the expansion*

$$\mu_k = \sum_{j=0}^{n+1-s} \Delta_j^{(0)} \rho_j^k \quad (k = 0, 1, \cdots, n)$$

let $\Delta_0^{(0)} = 1$. Then the variable elements Δ_j in the segment

$$\{\mu_i(\Delta_1, \Delta_2, \cdots, \Delta_{n+1-s})\}_{i=0}^n$$

are independent for all interior points of $\{M\}_{n+1-s}$.

We first show that $Q_n(x)$ cannot be continued into this region. Let $(\overset{\pm}{\sigma}_i)_1^s$ be the distribution of $Q_n(x)$. A necessary and sufficient condition for a continuation to be possible is that there are nonzero numbers $(\epsilon_i)_1^s$ and $(y_i)_1^{m-1}$, satisfying

$$\sum_{i=1}^{s} (\delta_i + \epsilon_i)\,\sigma_i^k = \sum_{j=1}^{n+1-s} (\Delta_j^{(0)} + y_j)\,\rho_j^k + \rho_0^k \quad (k = 0, 1, \cdots, n),$$

or, equivalently,

$$\sum_{i=1}^{s} \epsilon_i \sigma_i^k = \sum_{j=1}^{n+1-s} y_j \rho_j^k \quad (k = 0, 1, \cdots, n).$$

We have a homogeneous system of $n+1$ equations in $n+1$ unknowns with a Vandermonde determinant. Therefore $\epsilon_i = 0$ and $y_j = 0$, and $Q_n(x)$ is not an extremal polynomial at any point of the region except the initial point. But any interior point of $\{M\}_{n+1-s}$ can equally well be taken as the initial point; to it there corresponds a segment with the same fictitious nodal structure, but with different faithful weights δ_i' and fictitious weights Δ_i'. The previous reasoning applies. Thus the variable elements Δ_j are independent.

COROLLARY. *The set of principal polynomials of the variable segment* $[\mu_i(\Delta_1, \Delta_2, \cdots, \Delta_{n+1-s})]_0^n$ *is a family of polynomials of passport* $[n, s, P]$ *depending on* $n+1-s$ *independent parameters. We denote this family by* $Q_n(x, \Delta_1, \Delta_2, \cdots, \Delta_{n+1-s})$.

REMARK. Segments of the form $[(\mu_{k_p})_0^{s-1}; (\Theta_i)_1^{n+1-s}]$ are called segments of *basis type*; segments of the form $[\mu_i(\Delta_1, \Delta_2, \cdots, \Delta_{n+1-s})]_{i=0}^n$ are called segments of *fictitious nodal type*.

The size of the $(n+1-s)$-dimensional region of invariance of the passport of a segment depends essentially on the initial segment in the construction used in Theorems 33 and 34, i.e. on the more or less complete determination of the family of extremal polynomials by the variable segment.

§3. Determined segment-functionals

In what follows we consider only segments of basis type and clarify the optimal conditions for choice of the bases. For this purpose we begin by considering a special property of certain bases, which we call the property of "invariance of decomposition." Let $(\mu_{k_p})_{p=0}^{s-1}$, the basis of a variable segment, have the following property: for every choice of numbers $(\lambda_i)_1^s$, where $0 \leqq \lambda_1 < \lambda_2 < \cdots < \lambda_s \leqq 1$, the system of equations in the unknowns x_i

$$\sum_{i=1}^{s} x_i \lambda_i^{k_p} = \mu_{k_p} \quad (p = 0, 1, \cdots, s-1), \tag{47}$$

yields solutions x_i which have fixed sign for each x_i ($x_i \neq 0$) independently of the choice of $(\lambda_i)_1^s$ Then, if we assign to each λ_i the sign of the corresponding x_i, we obtain a distribution $(\overset{\pm}{\lambda_i})_1^s$ whose passport $[n, s, P]$ is the same for all choices of $(\lambda_i)_1^s$. When this special property occurs, we call the basis $(\mu_{k_p})_{p=0}^{s-1}$ a *basis with invariant decomposition*.

Examples 33 and 34 illustrate invariance of decomposition. In fact, the segment $0_0, 0_1, \cdots, 0_{n-2}, 1_{n-1}, \Theta$ has basis $0_0, \cdots, 0_{n-2}, 1_{n-1}$, which, decomposed according to arbitrary nodes $(\lambda_i)_1^n$, where $0 \leqq \lambda_1 < \lambda_2 < \cdots < \lambda_n \leqq 1$, has weights of alternating sign: $\overset{+}{\lambda}_n, \overset{-}{\lambda}_{n-1}, \cdots$, etc. Consequently the basis has invariant decomposition with characteristic $p = 0$. Similarly the segment $1, \rho, \rho^2, \cdots, \rho^{n-1}, \rho^n + \Theta$ with $\rho > 1$ has the same property (p. 70).

Theorem 35. *If there is a polynomial $Q_n(x)$ of passport $[n, s, P]$, and if the segment has a basis $[(\mu_{k_p})_0^{s-1}; (\Theta_i)_1^{n+1-s}]$ for which the number of nodes is at least s for all Θ_i, and if the basis has invariant decomposition with characteristic P, then the segment constructed with variable elements Θ_i determines, in the $l = (n+1-s)$-dimensional region $\{M\}_l$ of invariant passport, the family of all polynomials with this passport.*

We call such a segment a *determining* segment.

We shall prove that the same family is determined by every polynomial $L_n(x)$ with the same passport.

In fact, let $(\overset{\pm}{\sigma_i})_1^s$ be the distribution of $L_n(x)$. We decompose the basis $(\mu_{k_p})_{p=0}^{s-1}$ according to the nodes σ_i. By the hypothesis of invariance of decomposition, the system

$$\sum_{i=1}^{s} x_i \sigma_i^{k_p} = \mu_{k_p} \quad (p = 0, 1, \cdots, s-1)$$

can be solved for $x_i = x_i^{(0)}$ as follows: $x_i^{(0)} > 0$ for $\overset{+}{\sigma}_i$; $x_i^{(0)} < 0$ for $\bar{\sigma}_i$, i.e. the characteristic P is preserved. Then the segment $(\nu_i)_0^n$, where $\nu_k = \sum_{i=1}^{s} x_i^{(0)} \sigma_i^k$ $(k = 0, 1, \cdots, n)$ is a value of the variable basis of the segment, in which the elements have values from the region of invariant passport; consequently the principal polynomial of $(\nu_i)_0^n$, which is $L_n(x)$, belongs to the family $Q_n(x, \Theta_1, \cdots, \Theta_{n+1-s})$.

Thus the family $Q_n(x, \Theta_1, \cdots, \Theta_{n+1-s})$ is complete, i.e. contains all polynomials with the given passport.

COROLLARY. *On the boundary of the region* $\{M\}_l$ *the number of nodes of the serving polynomial is increased, i.e. the initial number s of nodes is a minimum for the whole l-dimensional space.*

In fact, the region $\{M\}_l$ is the complete region of invariance of the passport $[n, s, P]$. When the passport changes, the number s has to change. Suppose that we have $s_0 < s$ at a boundary point of $\{M\}_l$, and that the extremal polynomial has nodes $(\sigma_i)_1^{s_0}$. Then for a basis we have

$$\mu_{k_p} = \sum_{i=1}^{s_0} \sigma_i \delta_i^{k_p} \quad (p = 0, 1, \cdots, s-1),$$

where there are more equations than unknowns; if we augment the numbers σ_i by $s - s_0$ arbitrary numbers σ_i' chosen from $[0,1]$, we obtain

$$\mu_{k_p} = \sum_{i=1}^{s} \delta_i \sigma_i^{k_p} + \sum_{i=1}^{s-s_0} \Delta_i \sigma_i'^{k_p} \quad (p = 0, 1, \cdots, s-1),$$

where all $\Delta_i = 0$. Therefore the system $[(\sigma_i), (\sigma_i')]$ violates the hypothesis that the decomposition of the basis is invariant. Therefore $s_0 > s$.

REMARK 1. For any passport $[n, s, 0]$ it is easy to construct a specific segment whose basis has the property of invariance of decomposition.

Consider the segment $0_0, 0_1, \cdots, 0_{s-2}, 1_{s-1}, \Theta_1, \Theta_2, \cdots, \Theta_{n+1-s}$, which has, by criterion 2 (p. 64), at least s nodes; in addition, its basis $0_0, \cdots, 0_{s-2}, 1_{s-1}$ has the property of invariance of decomposition with respect to any $(\lambda_i)_1^s$ whose terms do not lie outside $[0,1]$, with characteristic $p = 0$. Consequently, this segment is determining for all polynomials of passport $[n, s, 0]$, if there exists at least one polynomial with this passport; but a polynomial of the form $T_n(\alpha x + \beta)$ is such a polynomial if α and β are chosen so that there are precisely s nodes on $[0,1]$.

REMARK 2. Polynomials of class II with all formally possible passports actually exist.

In fact, consider the passport $[n, s, p]$ with $s + p \leqq n + 1$ and $s \geqq \frac{1}{2}n + 2$, with arbitrary fixed n.

1. If for a given n there are polynomials with all passports, then for $n + 1$ there are polynomials of all passports of the form $[n + 1, s, P]$ with $s \geqq \frac{1}{2}n + 2$ and $s + p \leqq n + 1$.

In fact, let $Q_{\underline{n}}(x)$ have passport $[n, s, P]$; using its complete distribution we construct the segment $\mu_0, \mu_1, \cdots, \mu_{n-1}, \mu_n\,(\mathrm{I})$, and form $\mu_0, \mu_1, \cdots,$

$\mu_{n-1}, \mu_n, \mu_{n+1}^* \pm h$ (II), where μ_{n+1}^* is the best continuation and h is positive and small.

By the theorem on continuous deformation, (II) is served by a polynomial $Q_{n+1}(x)$ of passport $[n+1, s, P]$.

It remains to establish for each s the existence of polynomials of degree $n+1$ with the maximum number of permanences.

2. We can show at once that for each n there are polynomials for which $s + p = n + 1$, i.e. we are concerned with the passports

$$[n, n, 1], [n, n-1, 2], \cdots, [n, s, n+1-s], \cdots, [n, \tfrac{1}{2}n+2, \tfrac{1}{2}n-1]$$

or

$$[n, \tfrac{1}{2}(n+3), \tfrac{1}{2}(n-1)]$$

(according as $n = 2m$ or $n = 2m-1$). We consider only the case of even n.

All these polynomials (if they exist) must be primitive (p. 81). For, let $P_n(x)$ belong to $[n, s, n+1-s]$; if it is semiprimitive, a polynomial $P_n(Ax)$ with $A > 1$ has at least $s+1$ nodes; but adding nodes does not decrease p, and we have $s+1+n+1-s = n+2$, more than assumed.

COROLLARY. *The smallest possible* $q_{\min} = 2$.

In fact, $q = s - 1 - p = 2s - n - 2$ and $q_{\min} = n + 4 - n - 2 = 2$. (When n is odd, $q_{\min} = 1$.)

Consider

$$T_n(x) = 1 - 2^{2^n - 1} \cdot \prod_{i=1}^{n/2-1} (x - \overset{\pm}{\tau}_i)^2 x \cdot (1 - x).$$

Using the independence of the parameters σ_i in

$$Q_n(x) = 1 - C_{\max} \prod_{i=1}^{n/2-1} (x - \sigma_i)^2 \cdot (1 - x)$$

and considering $T_n(x)$ as a special case of $Q_n(x)$, we replace one of the τ_k by $\tau_k \pm \epsilon$ and then take $C_{\max} < 2^{2^n - 1}$ so that reducibility is preserved; the resulting polynomial has passport $[n, \tfrac{1}{2}n+2, \tfrac{1}{2}n-1]$.

Similarly, by moving two, three, etc. nodes of $T_n(x)$, we obtain polynomials with the passports listed above and characteristic $p_{\max}$.

CHAPTER IV

POLYNOMIALS OF PASSPORT $[n, n, 0]$

§1. Determining segments for $[n, n, 0]$ polynomials

The set $\{Q_n(x)\}$ of polynomials of class II (p. 37) is also the set of polynomials having all distributions of degree n with $s > \frac{1}{2}n + 1$.

So far, we know only the polynomials $\pm T_n(x)$ from this set—these are the only polynomials of passport $[n, n+1, 0]$. However, we note that if we know the analytic form of a polynomial $Q_n(x)$ from the set, we can immediately obtain a whole family $Q_n(\alpha x + \beta)$ of such polynomials, where $0 \leqq \alpha x + \beta \leqq 1$ for $0 \leqq x \leqq 1$. We shall call this construction a *transformation*; it is easily shown that we must have $|\alpha| \leqq 1$ and $0 \leqq \beta \leqq 1$.

DEFINITION. A polynomial $Q_n(x)$ of class II is called *primitive* if it satisfies the following boundary conditions:

$$|Q_n(0)| = 1; \quad |Q_n(0 - \epsilon)| > 1;$$
$$|Q_n(1)| = 1; \quad |Q_n(1 + \epsilon)| > 1$$

for all sufficiently small $\epsilon > 0$.

Typical examples of primitive polynomials are $\pm T_n(x)$.

A polynomial that fails to satisfy one of the hypotheses of primitivity is called *imprimitive*; it is obvious that it must satisfy either $|(Q_n(0 - \epsilon)| < 1$, or $|Q_n(1 + \epsilon)| < 1$, or both of these inequalities.

THEOREM 37. *If the polynomial $P_n(x)$ of a distribution is imprimitive, there is a linear mapping $Ax + B$ such that $Q_n(x) = P_n(Ax + B)$ is primitive.*

In fact, let us find the points $x_1 < 0$ and $x_2 > 0$ closest to $[0, 1]$ at which, with $\epsilon > 0$,

$$|P_n(x_1)| = 1; \quad |P_n(x_1 - \epsilon)| > 1; \quad |P_n(x_2)| = 1; \quad |P_n(x_2 + \epsilon)| > 1,$$

then $\max_{[x_1, x_2]} |P_n(x)| = 1$. Let the mapping $Ax + B$ take 0 and 1 to x_1 and x_2; then $Q_n(x) = P_n[(x_2 - x_1)x + x_1]$ is primitive ($Ax + B$ is not a transformation as defined above).

COROLLARY. *The mapping of $P_n(x)$ into the primitive polynomial $Q_n(x)$ extends its distribution by either one or two nodes (not less).*

It is convenient to single out among the imprimitive polynomials those for which the conditions of primitivity are satisfied at one end of the interval. We call such polynomials *semiprimitive*. To change a semiprimitive polynomial $P_n(x)$ into a primitive one, we can use one of the mappings $P_n(Ax)$ or $P_n(Ax + 1 - A)$, where $A > 1$, and then the distribution of the polynomial is extended by one node (not less).

If two polynomials $Q_n(x)$ and $P_n(x)$ are such that $Q_n(x) = P_n(\alpha x + \beta)$, where $\alpha x + \beta$ is a transformation, we call $Q_n(x)$ a transform of $P_n(x)$.

We note another consequence of Theorem 37 and our definitions.

THEOREM 38. *All polynomials of passport* $[n, n, 0]$ *or* $[n, n, 1]$ *are either primitive or else semiprimitive Čebyšev transformations.*

In fact, let a polynomial $P_n(x)$ of class II be different from $\pm T_n(x)$, and let its distribution contain $s = n$ nodes. Then $P_n(x)$ is necessarily at least semiprimitive, since otherwise it would gain two additional nodes under a linear mapping, i.e. it would become a polynomial of degree n with $n + 2$ nodes. If $P_n(x)$ is semiprimitive, we can find $Ax + B$ such that $P_n(Ax + B)$ is primitive and has $n + 1$ nodes, and then we necessarily have $P_n(Ax + B) \equiv \pm T_n(x)$. Hence $P_n(x) \equiv T_n(\alpha x + \beta)$, where $\alpha x + \beta$ is a transformation.

According to §2 of Chapter II, if there is at least one polynomial of passport $[n, n, 0]$, then the polynomials with this passport form a family of polynomials with one variable parameter of deformation. In addition, by Theorems 33 and 34, if these polynomials exist we can construct a determining segment for them: either 1) a basis segment with n zeros, or 2) a basis segment with a single fictitious node $\rho > 1$ (this is the simplest possible such segment).

Finally, since either of these two segments satisfies the condition of invariant decomposition with respect to n nodes (Theorem 35), each such segment $(\mu_i)_0^n$ together with its opposite $(-\mu_i)_0^n$ completely determines the whole family of polynomials of passport $[n, n, 0]$.

The question of the existence of such polynomials has a positive answer provided by the family of Čebyšev transformations $T_n(\alpha x)$, which have the required passport when $\alpha < 1$ and α is close to unity.

In studying the polynomials of passport $[n, n, 0]$ we shall need the following segments (with bases):

1. Consider the segment $(\mu_i)_0^n = -1, -\rho, -\rho^2, \cdots, -\rho^{n-1}, -\rho^n + \overline{0}$, where $\rho > 1$. We have already studied this segment in Examples 20

and 32. We recall the following results: the endpoints of the critical interval are

$$\mu_n' = -\rho^n + \bar{\Theta}' = -\rho^n + \frac{R_{T_n}(\rho)}{\rho} \quad \text{and} \quad \mu_n'' = -\rho^n + \frac{R_{T_n}(\rho)}{\rho - 1},$$

where $R_{T_n}(x) = \prod_{i=0}^{n}(x - \tau_i)$ is the resolvent of $T_n(x)$. The focus is determined from the resolvent $R_{T_{n-1}}(x)$ by the condition $-R_{T_{n-1}}(\rho) + \bar{\Theta} = 0$, whence $\bar{\Theta}^* = R_{T_{n-1}}(\rho)$. This follows because the number s of nodes of the segment $(\mu_i)_0^n$ cannot be less than n for any $\bar{\Theta}$, and consequently when $\bar{\Theta} = \bar{\Theta}^*$ the segment is served by one of the polynomials $\pm T_{n-1}(x)$.

At the left-hand end of the critical interval the weight δ_0 of the node $\tau_0 = 0$ becomes zero; and at the right-hand end the weight δ_n of the node $\tau_n = 1$ becomes zero. If we decompose $(\mu_i)_0^{n-1}$ with arbitrary n in terms of the nodes $(\lambda_i)_1^n$ $(0 \leqq \lambda_i \leqq 1)$ we obtain alternating signs for the weights: $\delta_n < 0$, $\delta_{n-1} > 0$, etc. Hence we have the following conclusion: If

$$\frac{R_{T_n}(\rho)}{\rho} < \bar{\Theta} < \frac{R_{T_n}(\rho)}{\rho - 1},$$

the segment cannot be served by a polynomial $\pm T_n(x)$, and consequently we have again confirmed the existence of polynomials of passport $[n, n, 0]$ which serve the segment in the critical interval, and indeed one for each point Θ and each one for some point. Since furthermore the condition of invariance of decomposition is satisfied up to the nth node, then (by Theorem 35) the segments $(+\mu_i)_0^n$ and $(-\mu_i)_0^n$ define all polynomials with this passport.

The family of polynomials determined by the segment $(\mu_i)_0^n$ in the critical interval (p. 82) will be denoted by $\bar{Q}_n(x, \bar{\Theta})$, and its distribution by $(\overset{\pm}{\sigma_i})_1^n$, where $\sigma_i = \sigma_i(\bar{\Theta})$. The weights $(\delta_i)_1^n$ of the nodes $(\sigma_i)_1^n$ in the decomposition

$$\mu_k = \sum_{i=1}^{n} \delta_i \sigma_i^k \quad (k = 0, 1, \cdots, n)$$

are

$$\delta_k = (-1)^{n-k+1} \cdot \frac{\prod (\rho - \sigma_i)}{\prod |\sigma_k - \sigma_i|}, \quad (i \neq k),$$

and $\bar{\Theta}$ and (σ) are connected by

$$\bar{\Theta} = \prod_{i=1}^{n} (\rho - \sigma_i). \tag{48}$$

The last equation is most simply obtained by considering the resultant $R_n(x) = \prod_{i=1}^{n}(x - \sigma_i)$ and taking account of the equation $R_n(\bar{\mu}) = 0$, i.e. $-\prod_1^n(\rho - \sigma_i) + \bar{\Theta} = 0$.

2. Consider the segment $(\nu_i)_0^n = 0_0, 0_1, \cdots, 0_{n-2}, -1_{n-1}, \Theta$. This was discussed in Examples 29 and 34.

The ends of the critical interval are $\Theta' = -\frac{1}{2}(n-1)$ and $\Theta'' = -\frac{1}{2}(n+1)$; at the left-hand end the weight δ_0 of the node $\tau_0 = 0$ becomes zero, and at the right-hand end, $\delta_n = 0$. The focus is $\Theta^* = -n/2$. For $-\frac{1}{2}(n+1) < \Theta < -\frac{1}{2}(n-1)$ the segment cannot be served by the polynomials $\pm T_n(x)$, and we have again confirmed the existence of polynomials of passport $[n, n, 0]$.

In addition, if we decompose the segment $(\nu_i)_0^{n-1}$ according to arbitrary $(\lambda_i)_1^n$, $0 \leqq \lambda_i \leqq 1$, we obtain alternating signs for the weights $(\delta_i)_1^n$: $\delta_n < 0, \cdots$. Since the condition of invariance of decomposition holds, the segment $(\nu_i)_0^n$ together with its opposite $(-\nu_i)_0^n$ determine *all* polynomials of passport $[n, n, 0]$.

The weights in the decomposition

$$\nu_k = \sum_{i=1}^{n} \delta_i \sigma_i^k \quad (k = 0, 1, \cdots, n)$$

are

$$\delta_k = \frac{(-1)^{n-k+1}}{\prod_{i \neq k} |\sigma_k - \sigma_i|};$$

the relation between Θ and (σ_i) is most easily found by observing that the resolvent $R_n(x) = \prod_1^n(x - \sigma_i)$ satisfies $R_n(\bar{\mu}) = 0$. Then we obtain

$$\Theta = -\sum_1^n \sigma_i. \tag{49}$$

The family of polynomials determined by segment 2 will be denoted by $(Q_n(x, \Theta))$.

Let us trace the deformation of the polynomials $Q_n(x, \Theta)$ and $\bar{Q}_n(x, \bar{\Theta})$ when Θ and $\bar{\Theta}$ decrease from the right-hand end of the critical interval to the left-hand end. In both cases we start from $+T_n(x)$, for which the right-hand node is $\tau_n = 1$; the remaining nodes have weights $(\delta_i)_1^n$ of alternating sign, and in both cases each weight is of constant sign throughout the critical interval (in fact $\delta_n < 0$; $\delta_{n-1} > 0$, etc.), since none of the n weights can become zero (n is the minimal number of faithful nodes for the segment).

In both cases, the extremal polynomial must become $-T_{n-1}(x)$ at the focus. Thus, in the right-hand part of the critical interval for both segments 1 and 2, the families of polynomials $\overline{Q}_n(x, \overline{\Theta})$ and $Q_n(x, \Theta)$ are identical.

It is clear that in the left-hand part of the critical interval segments 1 and 2 are respectively served by the polynomials $(-1)^{n-1}\overline{Q}_n(1-x, \overline{\Theta})$ and $(-1)^{n-1}Q_n(1-x, \Theta)$, and consequently the families are again identical. Thus for a complete investigation of the polynomials of passport $[n, n, 0]$ it is enough to investigate the polynomials corresponding to one of the segments, and only in the right-hand part of the critical interval. We shall consider segment 2, so that we are investigating the polynomials $Q_n(x, \Theta)$ with $-\frac{1}{2}n \leqq \Theta \leqq -\frac{1}{2}(n-1)$.

THEOREM 39. *The polynomials $Q_n(x, \Theta)$ change their analytic form at the point*

$$\Theta = -\frac{n-1}{2\cos^2(\pi/2n)}.$$

In fact, $Q_n(x, \Theta) \equiv T_n(\alpha x)$, where $\Theta = -(n-1)/2\alpha$, for

$$-\frac{n-1}{2\cos^2(\pi/2n)} \leqq \Theta \leqq -\frac{n-1}{2};$$

but for

$$-\frac{n}{2} < \Theta < -\frac{n-1}{2\cos^2(\pi/2n)}$$

the polynomials $Q_n(x, \Theta)$ form a family of new primitive polynomials.

Indeed, we found that at $\Theta' = -\frac{1}{2}(n-1)$ the polynomial $T_n(x)$ loses its weight at the node $\tau_n = 1$, but preserves the signs of the weights at the other nodes. Consequently by Theorem 27 (on continuous deformation) the segment is served by a polynomial $T_n(\alpha x)$ with $\alpha > 0$.

The distribution of $T_n(\alpha x)$ is

$$0 = \frac{\tau_0}{\alpha} < \frac{\tau_1}{\alpha} < \cdots < \frac{\tau_{n-2}}{\alpha} < \frac{\tau_{n-1}}{\alpha}.$$

Since $\tau_{n-1} = \cos^2(\pi/2n)$, a necessary and sufficient condition that $s = n$ is that $\cos^2(\pi/2n) \leqq \alpha < 1$, and this is the exact interval in which $T_n(\alpha x)$ serves.

We now determine the relation between α and Θ. To do this, we decompose the segment $(\mu_i)_0^{n-1} = 0_0, \cdots, 0_{n-2}, -1_{n-1}$ in terms of the n nodes of $T_n(\alpha x)$. Denoting the corresponding weights by Δ_i, we have

$$\mu_k = \frac{1}{\alpha_k} \sum_{i=0}^{n-1} \Delta_i \tau_i^k.$$

But $\sum_{i=0}^{n-1} \Delta_i \tau_i^k$ is the decomposition of $0_0, \cdots, 0_{n-2}, -\alpha^{n-1}, \alpha^n \Theta$ in terms of the nodes of $T_n(x)$, with $\tau_n = 1$ omitted. Consequently $-\frac{1}{2}(n-1) = \alpha\Theta$, i.e. $\Theta = -\frac{1}{2}(n-1)/(2\alpha)$.

Therefore the exact interval of applicability of the polynomials $T_n(\alpha x)$ is

$$-\frac{n-1}{2\cos^2(\pi/2n)} \leqq \Theta < -\frac{n-1}{2}. \tag{50}$$

But in the interval

$$-\frac{n}{2} < \Theta < -\frac{n-1}{2\cos^2(\pi/2n)} \tag{51}$$

the segment $(\mu_i)_0^n$ cannot be served by any other Čebyšev semiprimitive transformation. In fact, such transformations must be $\pm T_n(\alpha x)$ or $\pm T_n[\alpha x + (1-\alpha)]$; but $+T_n(\alpha x)$ has already been used, $-T_n(\alpha x)$ is unsuitable because of its sequence of changes of sign, and $\pm T_n[\alpha x + (1-\alpha)]$ has the fixed node 1, which is impossible for the segment in the interval under consideration. Thus in the interval (51) the segment $(\mu_i)_0^n$ is served by polynomials which are not semiprimitive transformations of $\pm T_n(x)$, and are therefore new primitive polynomials (p. 82). We denote them by $Z_n(x, \Theta)$.

Let us consider the determining segments of Čebyšev semiprimitive transformations.

We consider the segment $(\mu_i)_0^n = -1_0, -1_1, \cdots, -1_{n-1}, -1+\Theta$, and determine the critical interval $(-1+\Theta', -1+\Theta'')$.

Clearly $\Theta' = \Theta^* = 0$.

To determine Θ'' we decompose the segment $(\nu_i)_0^n = 0_0, \cdots, 0_{n-1}, \Theta$ in terms of the nodes $(\tau_i)_0^n$, and then

$$\delta_k = (-1)^{n-k} \frac{\Theta}{\prod_{i \neq k} |\tau_k - \tau_i|} \quad (k = 0, 1, \cdots, n).$$

In the system of equations $\sum_{i=0}^n \delta_i \tau_i^k = \nu_k$ $(k = 0, 1, \cdots, n)$, we transpose $\delta_n \tau_n^k = \delta_n$ to the right and put $\Theta = \prod_0^{n-1}(1-\tau_i)$; we then obtain the decomposition of the segment

$$-1, -1, \cdots, -1_{n-1}, \left[-1 + \prod_0^{n-1}(1-\tau_i)\right]$$

in terms of the nodes $(\tau_i)_0^{n-1}$ and consequently $\Theta'' = \prod_{i \neq n}(1-\tau_i)$.

At points of the critical interval close to Θ'' the passport of the segment is $[n,n,0]$; the faithful distribution $(\overset{\pm}{\sigma}_i)_1^n$ of the segment has the fixed node $\sigma_1 = 0$, and $\sigma_n < 1$. Since there is a family of polynomials, in fact $T_n(\alpha x)$, which serves this segment for

$$\cos^2(\pi/2n) \leqq \alpha < 1, \tag{52}$$

it follows that all these polynomials are extremal for $(\mu_i)_0^n$. It remains to show that, in the critical interval, all extremal polynomials belong to the family $T_n(\alpha x)$. To do this we have to establish a one-to-one correspondence between Θ and α.

We have already obtained the relation between them,

$$\Theta = \prod_{i=0}^{n-1} \left(1 - \frac{\tau_i}{\alpha}\right),$$

which shows that $\Theta(\alpha)$ is a continuous monotonic function that increases with α; for $\alpha = 1$, $\Theta = \Theta''$; for $\alpha = \tau_{n-1} = \cos^2(\pi/2n)$, $\Theta = \Theta^* = 0$. Therefore $T_n(\alpha x)$ serves the segment for all points of the critical interval.

REMARK. The segment $(\nu_i)_0^n = -1_0, 0_1, \cdots, 0_{n-1}, (-1)^n\Theta$ is a determining segment for the polynomials $T_n[\alpha(1-x)]$ when $\cos^2(\pi/2n) \leqq \alpha < 1$, since $\nu_i = \mu_{0,i}$. The segments $1_0, 1_1, \cdots, 1_{n-1}, 1-\Theta$ and $1_0, 0_1, \cdots, 0_{n-1}, (-1)^{n-1}\Theta$ respectively determine all polynomials of the forms $-T_n(\alpha x)$ and $-T_n[\alpha(1-x)]$, and only these (in the critical interval). Consequently, since every semiprimitive polynomial of passport $[n,n,0]$ belongs to one of the transformations

$$\pm T_n(\alpha x); \quad \pm T_n[\alpha(1-x)],$$

it follows that

$$T_n(\alpha x + 1 - \alpha) = (-1)^n T_n[\alpha(1-x)]; \quad T_n(-\alpha x + 1) = (-1)^n T_n(\alpha x).$$

We return to the segment $(\mu_i)_0^n = 0_0, 0_1, \cdots, 0_{n-2}, -1_{n-1}, \Theta$.

We have shown that in the interval

$$-\frac{n}{2} < \Theta < -\frac{n-1}{2\cos^2(\pi/2n)}$$

the segment defines new polynomials that are not transformations of $T_n(x)$. This family of polynomials $Q_n(x,\Theta)$ of passport $[n,n,0]$ will be denoted by $Z_n(x,\Theta)$.

We have

$$\lim_{\Theta \to -n/2+} z_n(x,\Theta) = -T_{n-1}(x).$$

We note, incidentally, that these new polynomials exist, as inequality (50) shows, starting with $n \geqq 3$ (for $n \leqq 2$ they are not needed and hence do not exist).

Let us consider the nature of the extremal polynomials for $(\mu_i)_0^n$ in the left-hand part of the critical interval $-\frac{1}{2}(n+1) \leqq \Theta \leqq -\frac{1}{2}n$. Since the polynomial $-T_n(x)$ loses its weight at the node $\tau_0 = 0$ when $\Theta = -\frac{1}{2}(n+1)$, the segment is served instead by the transformation

$$(-1)^{n-1}T_n[\alpha(1-x)] = -T_n(\alpha x + 1 - \alpha).$$

We now find the exact interval where these polynomials serve, and the relation between α and Θ.

For the segment $(\mu_{0,i})_0^n = 0_0, \cdots, 0_{n-2}, (-1)^n, (-1)^n(n+\Theta)$ we have $T_n[\alpha(1-x)]$ as extremal polynomial for

$$-\frac{n-1}{2\cos^2(\pi/2n)} < \Theta < -\frac{n-1}{2}$$

and the same for the segment $0_0, \cdots, 0_{n-2}, (-1)^n, (-1)^n\Theta$ in the interval

$$n - \frac{n-1}{2\cos^2(\pi/2n)} < \Theta < \frac{n-1}{2};$$

hence when n is odd we find, replacing Θ by $-\Theta$, that the segment $0_0, \cdots, 0_{n-2}, -1, \Theta$ is served by the polynomials $-T_n[\alpha(1-x)]$ precisely in the interval

$$-\frac{n+1}{2} < \Theta \leqq -n + \frac{n-1}{2\cos^2(\pi/2n)}.$$

When n is even we find in a similar way, by using the segment $(-\mu_{0,i})_0^n$ and replacing $-\Theta$ by Θ, the segment $0_0, \cdots, 0_{n-2}, -1, \Theta$, which is served by the polynomials $-T_n[\alpha(1-x)]$ precisely in the interval

$$-\frac{n+1}{2} < \Theta \leqq -n + \frac{n-1}{2\cos^2(\pi/2n)}.$$

The relation between α and Θ in the left-hand part of the critical interval is clearly

$$\alpha = \frac{n-1}{2(n+\Theta)},$$

where Θ has its original value.

Thus, finally, for the two opposite segments $(\mu_i)_0^n$ and $(-\mu_i)_0^n$ the critical interval falls into subintervals with serving polynomials as follows:

Endpoints of Subintervals

$$-\frac{n+1}{2};\quad \left(-n+\frac{n-1}{2\cos^2(\pi/2n)}\right);\quad -\frac{n}{2};\quad -\frac{n-1}{2\cos^2(\pi/2n)};\quad -\frac{n-1}{2}.$$

Serving Polynomials

$$\underbrace{(-1)^{n-1}T_n[\alpha(1-x)]}\ \underbrace{(-1)^{n-1}Z_n(1-x,\Theta)}\ \underbrace{Z_n(x,\Theta)}\ \underbrace{T_n(\alpha x)}$$

$$\alpha=\frac{n-1}{2(n+\Theta)}\quad \text{New Polynomials}\quad \alpha=-\frac{n-1}{2\Theta}.$$

Endpoints of Subintervals

$$\frac{n-1}{2};\quad \frac{n-1}{2\cos^2(\pi/2n)};\quad \frac{n}{2};\quad \left(n-\frac{n-1}{2\cos^2(\pi/2n)}\right);\quad \frac{n+1}{2}.$$

Serving Polynomials

$$\underbrace{-T_n(\alpha x)}\ \underbrace{-Z_n(x,\Theta)}\ \underbrace{(-1)^nZ_n(1-x,\Theta)}\ \underbrace{(-1)^nT_n[\alpha(1-x)]}$$

$$\alpha=-\frac{n-1}{2\Theta}\quad \text{New Polynomials}\quad \alpha=\frac{n-1}{2(n+\Theta)}.$$

A graph of the deformation of the polynomials for the family $[n,n,0]$ is shown in Figure 1 for $n=6$; it makes use of Theorem 43.[1)]

It follows from the preceding calculations that as n increases the polynomials $\pm Z_n(x,\Theta)$ and $\pm Z_n(1-x,\Theta)$ tend to crowd out the Čebyšev transformations; in fact, the length of the critical interval for either of the two determining segments is $\Theta''=\Theta'=1$; the length of the interval where the polynomials Z_n serve tends to

$$\lim_{n\to\infty}\left(n-\frac{n-1}{\cos^2(\pi/2n)}\right)=1,$$

and hence it is clear that the distance between τ_{n-1} and $\tau_n=1$ tends to zero as $n\to\infty$.

The segment $(\mu_i)_0^n=-1,-\rho,-\rho^2,\cdots,-\rho^{n-1},-\rho^n+\bar\Theta$ with constant $\rho>1$ yields a similar deformation of its extremal polynomials.

In the critical interval we shall determine the parameter $\bar\Theta$ of the

---

1) For simplicity the graphs of polynomials on $[0,1]$ are conventionally represented by broken lines.

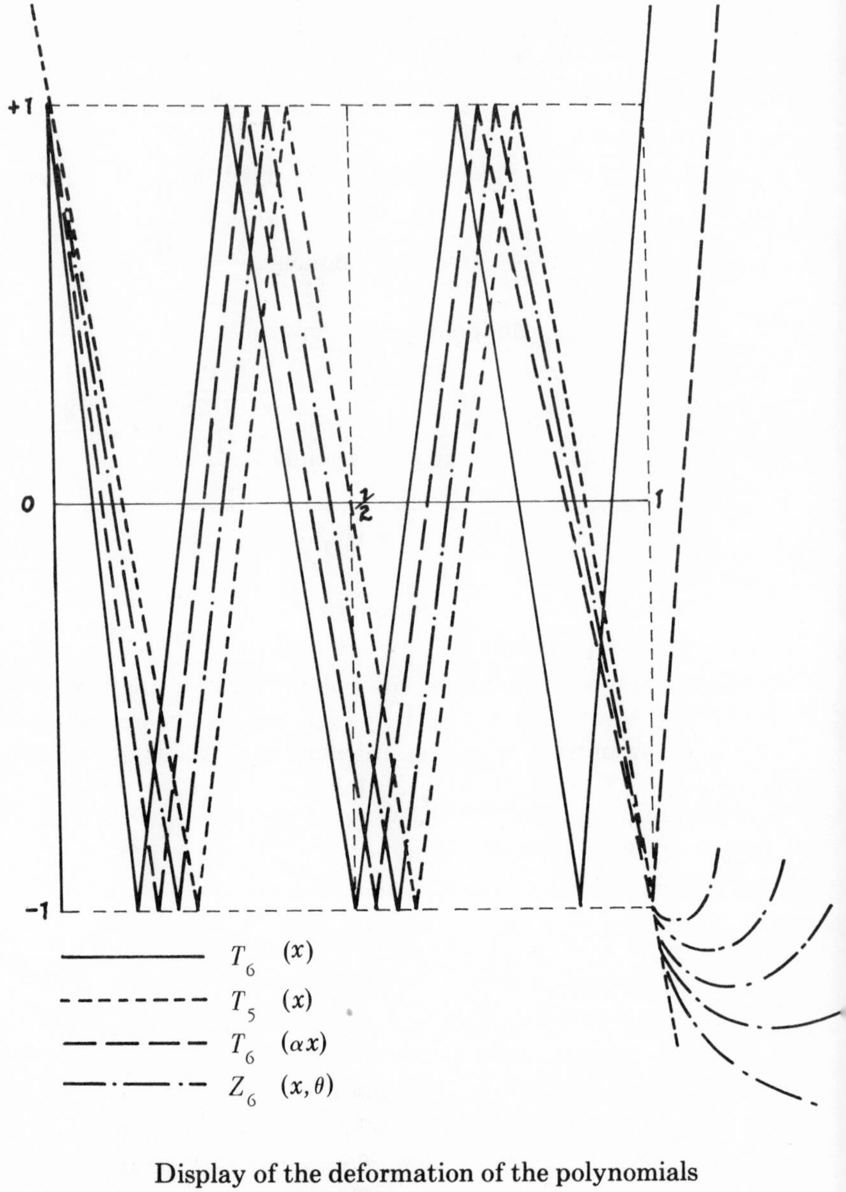

Display of the deformation of the polynomials
$Q_6(x, \Theta)$ for $\Theta^* \leqq \Theta \leqq \Theta''$

FIGURE 1

boundaries of the subintervals in which Čebyšev transformations serve. We have obtained the equation $\overline{\Theta} = \prod_1^n(\rho - \sigma_i)$, where $(\sigma_i)_1^n$ are the faithful nodes of the segment in the critical interval. Consequently, the lower bound of values for which $T_n(\alpha x)$ serves is

$$\overline{\Theta} = \prod_{i=0}^{n-1}\left(\rho - \frac{\tau_i}{2\cos^2(\pi/2n)}\right),$$

and the upper bound is

$$\overline{\Theta} = \prod_{i=1}^{n}\left(\rho - 1 + \frac{\tau_i}{2\cos^2(\pi/2n)}\right)$$

for the polynomials $(-1)^n T_n[\alpha(1-x)]$. Thus in the interval

$$R_{T_{n-1}}(\rho) < \overline{\Theta} < \prod_0^n\left(\rho - \frac{\tau_i}{2\cos^2(\pi/2n)}\right)$$

the segment defines a family of polynomials $Z_n(x, \overline{\Theta})$ which is the same as the family $Z_n(x, \Theta)$ (the relation between Θ and $\overline{\Theta}$ is given in Example 36), and in the interval

$$\prod_{i=0}^{n}\left(\rho - 1 + \frac{\tau_i}{2\cos^2(\pi/2n)}\right) < \overline{\Theta} < R_{T_{n-1}}(\rho)$$

it defines the family of polynomials $(-1)^n Z_n(1-x, \overline{\Theta})$.

§2. Analytic construction of $[n, n, 0]$ polynomials

In discussing the analytic construction of the polynomials $Z_n(x, \Theta)$ defined by the segment $0_0, \cdots, 0_{n-2}, -1_{n-1}, \Theta$ in the interval

$$-\frac{n}{2} < \Theta < -\frac{n-1}{2\cos^2(\pi/2n)},$$

we need the concept of equivalent segments.

DEFINITION. Two segments $(\mu_i)_0^n$ and $(\nu_i)_0^n$, each containing a variable element, say $\mu_k = \Theta$ and $\nu_l = \vartheta$ $(k > 0,\ l > 0)$ are called *equivalent* in the respective intervals $\alpha \leqq \Theta \leqq \beta$, $\gamma \leqq \vartheta \leqq \delta$ if they define the same family of polynomials of constant passport; this means that if $X_n(x, \Theta)$ and $Y_n(x, \vartheta)$ are the families of polynomials determined by the given segments, and $\{\sigma_i(\Theta)\}_1^s$ and $\{\rho_i(\vartheta)\}_1^s$ are their nodes, then Θ and ϑ are connected by the one-to-one relations $\vartheta = \phi(\Theta)$, $\Theta = \psi(\vartheta)$ in such a way that $\sigma_i[\psi(\vartheta)] = \rho_i(\vartheta)$; $\rho_i[\phi(\Theta)] = \sigma_i(\Theta)$;

$$Y_n[x, \phi(\Theta)] \equiv X_n(x, \Theta);$$
$$X_n[x, \psi(\vartheta)] \equiv Y_n(x, \vartheta).$$

EXAMPLE 36. The segments $(\mu_i)_0^n = 0_0, 0_1, \cdots, 0_{n-2}, -1_{n-1}, \Theta$ and $(\nu_i)_0^n = -1, -\rho, \cdots, -\rho^{n-1}, -\rho^n + \vartheta$ are equivalent in the intervals

$$-\frac{n}{2} < \Theta < -\frac{n-1}{2\cos^2(\pi/2n)}$$

and

$$-\rho^n + \prod_{i=0}^{n-1}(\rho - \tau_i') < \vartheta < -\rho^n + \prod_{i=0}^{n-1}\left(\rho - \frac{\tau_i}{\cos^2(\pi/2n)}\right)$$

(as has already been proved).

The relation between Θ and ϑ can be found as follows. The segment $(\mu_i)_0^n$ defines the family $Z_n(x, \Theta) = \sum_0^n \beta_i(\Theta) x^i$. In the segment $(\nu_i)_0^n$ we take $\vartheta = \text{const}$. For this segment we find the extremal polynomial, which, by what has been proved, is an element of the family $Z_n(x, \Theta)$. Then the extremal polynomial must satisfy $Z_n(\bar{\nu}, \Theta) = -Z_n(\rho, \Theta) + \beta_n(\Theta)\vartheta = \max \Theta$. Consequently

$$\vartheta = \frac{1}{\beta_n'(\Theta)} \cdot \frac{\partial Z_n(\rho, \Theta)}{\partial \Theta}.$$

(The differentiability of Z_n with respect to the parameter will be discussed below.)

REMARK. Each of the polynomials

$$R_n(x, \Theta) = \prod_1^n [x - \sigma_i(\Theta)] = \sum_{i=0}^n \alpha_i(\Theta) x^i;$$

$$\overline{R}_n(x, \vartheta) = \prod_{i=1}^n [x - \rho_i(\vartheta)] = \sum_{i=0}^n \overline{\alpha}_i(\vartheta) x^i$$

is the resolvent of the corresponding polynomial X_n of Y_n and of the corresponding segment $(\mu_i)_0^n$ or $(\nu_i)_0^n$. The fundamental property of the resolvents is $R_n(\bar{\mu}, \Theta) = 0$, $R_n(\bar{\nu}, \vartheta) = 0$. Since the segments are equivalent, it follows that $\overline{\alpha}_i(\vartheta) \equiv \alpha_i(\Theta)$ for $\vartheta = \phi(\Theta)$, and $R_n(x, \Theta) \equiv \overline{R}_n(x, \vartheta)$.

Consequently both $R_n(\bar{\nu}, \Theta) = 0$ and $\overline{R}_n(\bar{\mu}, \vartheta) = 0$.

If the nodes of the polynomial $Z_n(x, \Theta)$ are denoted by $(\sigma_i)_1^n$, then $\sigma_1 = 0$; $\sigma_n = 1$; and the resolvent of this polynomial (or of the segment (μ)) has the form

$$R_n(x,\Theta) = \alpha_i(\Theta)x + \cdots + \alpha_{n-1}(\Theta)x^{n-1} + x^n$$

for

$$-\frac{n}{2} < \Theta < -\frac{n-1}{2\cos^2(\pi/2n)}.$$

THEOREM 40. *The system of segments*

$$(53)\qquad \begin{aligned} (\mu_i)_0^n &= 0_0, 0_1, \cdots, 0_{n-2}, -1_{n-1}, \Theta; \\ (\mu_i^{(1)})_0^n &= 0_0, 0_1, \cdots, +1_{n-2}, 0_{n-1}, \Theta^{(1)}; \\ &\cdots\cdots\cdots\cdots\cdots \\ (\mu_i^{(n-2)})_0^n &= 0_0, (-1)^{n-1}, \cdots, 0_{n-2}, 0_{n-1}, \Theta^{(n-2)} \end{aligned}$$

is a system of equivalent segments for the family of polynomials $Z_n(x,\Theta)$, *determined by the first segment; if we put* $Z_n(x,\Theta) = \sum_0^n \beta_i(\Theta)x^i$, *then the variables* $\Theta^{(1)}, \cdots, \Theta^{(n-2)}$ *are connected with* Θ *by*

$$(54)\qquad \Theta^{(n-m-1)} = (-1)^{n-m}\frac{\beta_m'(\Theta)}{\beta_n'(\Theta)} \quad (m = 1, 2, \cdots, n-2).$$

The coefficients $[\alpha_i(\Theta)]_1^{n-1}$ of the resolvent $R_n(x,\Theta)$ of the polynomial $Z_n(x,\Theta)$ are given by

$$(55)\qquad \alpha_{n-1} = \Theta,\ \alpha_{n-2} = \frac{\beta_{n-2}'(\Theta)}{\beta_n'(\Theta)}, \cdots, \alpha_1 = \frac{\beta_1'(\Theta)}{\beta_n'(\Theta)}.$$

We now establish the equivalence of the segments (53), and find the relations between their variables.

From what we have already said, it is clear that these segments all have passport $[n, n, 0]$[2] (in the critical intervals). In addition each of them has the property of invariance of decomposition, and consequently also the property of complete determination of the family of polynomials of that passport.

For any such segment

$$0_0, 0_1, \cdots, 0_{m-1}, (-1)^{n-m}, 0_{m+1}, \cdots, 0_{n-1}, \Theta \quad (m = \overline{1, 2, \cdots, n-1})$$

the endpoints of the critical interval for Θ can be found from the material of §4, Chapter I. In fact, if we decompose the segment according to the nodes of $T_n(x)$, we obtain the weights

[2] With the same alternations of sign.

$$\delta_k = (-1)^{n-k} \frac{[\Theta + s_{n-m}^{(k)}]}{\prod_{i \neq k} |\tau_k - \tau_i|} \quad (k = 0, 1, \cdots, n)$$

(the value of $s_{n-m}^{(k)}$ is given on p. 47); putting $\delta_k = 0$, we have $\Theta_{\max} = \Theta'' = -s_{n-m}^{(n)}$; $\Theta_{\min} = \Theta' = -s_{(n-m)}^{(0)}$.

Therefore at the endpoint $\Theta = \Theta''$ the node $\tau_n = 1$ disappears and $T_n(x)$ is replaced by $T_n(\alpha x)$ with $\cos^2(\pi/2n) < \alpha < 1$, and then $T_n(\alpha x)$ is replaced by $Z_n(x, \Theta)$. The focus Θ^* of the segment can be found from the resolvent $R_{T_{n-1}}(x)$ and is $(-1)^{n-m-1}\alpha'_m$, where α'_m is the coefficient of x^m in the resolvent.

To express $\Theta^{(m)}$ in terms of Θ we proceed in the same way as in Example 36. We have $Z_n(\bar{\mu}^{(m)}, \Theta) = \max$ or $\beta_n(\Theta)\Theta^{(m)} + (-1)^{m+1}\beta_{n-m-1}(\Theta) = \max \Theta$, whence we immediately obtain

$$\Theta^{(m)} = (-1)^m \frac{\beta'_{n-m-1}(\Theta)}{\beta'_n(\Theta)},$$

i.e. formula (54). To find the coefficients $\alpha_i(\Theta)$ of the resolvent of $Z_n(x_1, \Theta)$ we observe that by general properties of the resolvent we have $R_n(\bar{\mu}^{(m)}, \Theta) = 0$, which leads to the following system of $n-1$ equations with nonvanishing determinant in the $n-1$ unknowns $(\alpha_i)_0^{n-1}$:

$$(-1)^m \alpha_{n-m}(\Theta) + \Theta^{(m-1)} = 0 \quad (m = 1, 2, \cdots, n-1),$$

whence

$$\alpha_{n-1}(\Theta) = \Theta; \ \alpha_{n-2} = \frac{\beta'_{n-2}(\Theta)}{\beta'_n(\Theta)}; \cdots; \alpha_1 = \frac{\beta'_1(\Theta)}{\beta'_n(\Theta)},$$

i.e. (55) holds.

This completes the proof of Theorem 40.

The polynomial $Z_n(x, \Theta)$ and its resolvent $R_n(x, \Theta)$ are connected by the following algebraic relation:

$$R_n(x, \Theta) \equiv x(x-1) \frac{\partial Z_n(x, \Theta)/\partial x}{n\beta_n(\Theta)x - \beta_1(\Theta)/\alpha_1(\Theta)} \tag{56}$$

(where $\alpha_1(\Theta)$, as a coefficient of the resolvent, cannot be zero).

In fact, each polynomial $Z_n(x, \Theta)$ has 0 and 1 as nodes, and the remaining nodes are the roots of its derivative, except for one root $\lambda > 1$. This is evident from the nature of the deformation of $Z_n(x, \Theta)$ which starts with $T_n(\cos^2(\pi/2n)x)$ and ends with $-T_{n-1}(x)$.

The root λ can be found from the coefficients α_i and β_i by means of the identity

$$x - \lambda \equiv \frac{\partial Z_n(x,\Theta) \cdot x(x-1)/\partial x}{n\beta_n(\Theta) \cdot R_n(x,\Theta)};$$

putting $x = 0$, we have

$$\lambda = \beta_1(\Theta)/n\alpha_1(\Theta)\beta_n(\Theta).$$

Formula (56) follows immediately.

To determine the analytic form of $Z_n(x,\Theta)$ it is convenient to express the deformation in terms of a different parameter.

THEOREM 41. *Denote the leading coefficient of* $Z_n(x,\Theta)$ *by* ϑ, *and take* ϑ *as parameter, so that* $Z_n(x,\Theta) = \zeta_n(x,\vartheta)$. *Then* $\zeta_n(x,\vartheta)$ *has the following properties: its coefficients are differentiable functions of* ϑ, *and the resolvent* $R_n(x,\vartheta)$ *is equal to its derivative with respect to* ϑ, *i.e.*

$$R_n(x,\vartheta) = \frac{\partial}{\partial\vartheta}\zeta_n(x,\vartheta). \tag{57}$$

We first show that the leading coefficient ϑ is a suitable parameter for the deformation. In fact, by Theorem 20 and Corollary 2 of Theorem 21, $Z_n(x,\Theta)$, as a single-valued function of Θ, has the property that its leading coefficient $\beta_n(\Theta)$ is a single-valued continuous function, increasing with respect to Θ in the critical interval (Θ', Θ''). Hence $\vartheta = \beta_n(\Theta)$ is in one-to-one correspondence with the family $Z_n(x,\Theta)$ and so ϑ can be taken as parameter for the family, i.e. $\zeta_n(x,\vartheta) \equiv Z_n(x,\Theta)$, where ϑ and Θ are inverse continuous monotonic functions. Put $\Theta = \psi(\vartheta)$ and

$$\zeta_n(x,\vartheta) = \vartheta x^n + y_{n-1}(\vartheta)x^{n-1} + \cdots + y_1(\vartheta)x + (-1)^n, \tag{58}$$

where $y_i(\vartheta) = \beta_i[\psi(\vartheta)]$.

We now show that $y_i(\vartheta)$ is differentiable with respect to ϑ.

The relation between ϑ and Θ can be obtained in a different way. In fact, putting $\Theta =$ const., we find a polynomial that serves the segment $0_0, \cdots, 0_{n-2}, -1_{n-1}, \Theta$ in the family $\zeta_n(x,\vartheta)$.

We have $\vartheta \cdot \Theta - y_{n-1}(\vartheta) = \max(\vartheta)$, and if we assume that y_{n-1} is differentiable, we can put $\Theta = y'_{n-1}(\vartheta)$; but $\Theta = \psi(\vartheta)$ is automatically continuous, so $y'_{n-1}(\vartheta)$ exists. Consider the segment $0_0, \cdots, 0_{n-3}, +1_{n-2}, 0_{n-1}, \Theta^{(1)}$. Since this segment defines, by Theorem 40, the same family of polynomials, we can find a polynomial that serves it in the family $\zeta_n(x, \vartheta)$. In just the same way we have $\vartheta\Theta^{(1)} + y_{n-2}(\vartheta) = \max(\vartheta)$, i.e if $y'_{n-2}(\vartheta)$ exists, $\Theta^{(1)} = -y'_{n-2}(\vartheta)$. But $\Theta^{(1)}$ is connected with ϑ by a by a single-valued continuous monotonic function, $\Theta^{(1)} = \psi_1(\vartheta)$, which consequently is $-y'_{n-2}(\vartheta)$. In the same way, by using the other segments of the system (53), we can establish the differentiability of the coefficients $y_i(\vartheta)$ $(i = n-1, n-2, \cdots, 1)$.

Thus $\zeta_n(x, \vartheta)$ is a differentiable function of the parameter.

Now let $R_n(x, \vartheta) = x^n + \alpha_{n-1}(\vartheta)x^{n-1} + \cdots + \alpha_1(\vartheta)x$ be the resolvent of $\zeta_n(x, \vartheta)$. By the properties of the resolvent we have $R_n(\bar{\mu}, \vartheta) = 0$; $R_n(\bar{\mu}^{(1)}, \vartheta) = 0; \cdots; R_n(\bar{\mu}^{(n-2)}, \vartheta) = 0$.

Consequently $\alpha_{n-1}(\vartheta) = y'_{n-1}(\vartheta)$; $\alpha_{n-2}(\vartheta) = y'_{n-2}(\vartheta); \cdots; \alpha_1(\vartheta) = y'_1(\vartheta)$; and finally

$$R_n(x, \vartheta) = \frac{\partial}{\partial\vartheta}\, \zeta_n(x, \vartheta).$$

This completes the proof of Theorem 41.

To obtain the system of differential equations that the $n-1$ functions $y_i(\vartheta)$ satisfy, we note that $R_n(x, \vartheta)$ and $\zeta_n(x, \vartheta)$ satisfy not only the differential relation (57) but also an algebraic relation similar to (56):

$$R_n(x, \vartheta) = \frac{\partial\zeta_n(x, \vartheta)/\partial x}{n\vartheta x - y_1(\vartheta)/y'_1(\vartheta)}\, x(x-1). \tag{59}$$

Here $y_1(\vartheta)$ is monotonic and $y'_1(\vartheta) \neq 0$.

Hence we obtain

$$x(x-1)\frac{\partial}{\partial x}\, \zeta_n(x, \vartheta) = \left(n\vartheta x - \frac{y_1(\vartheta)}{y'_1(\vartheta)}\right)\frac{\partial}{\partial\vartheta}\, \zeta_n(x, \vartheta), \tag{60}$$

a system of ordinary first-order differential equations for $y_1(\vartheta), y_2(\vartheta), \cdots, y_{n-1}(\vartheta)$.

If we compare coefficients of x^i $(i = 1, 2, \cdots, n-1)$, we can write the system in expanded form as

$$\begin{aligned}(x-1)\,[n\vartheta x^n &+ (n-1)y_{n-1}(\vartheta)x^{n-1} \\ &+ (n-2)y_{n-2}(\vartheta)x^{n-2} + \cdots + y_1(\vartheta)x] \\ = \left[n\vartheta x - \frac{y_1(\vartheta)}{y'_1(\vartheta)}\right]&[x^n + y'_{n-1}(\vartheta)x^{n-1} + y'_{n-2}(\vartheta)x^{n-2} + \cdots + y'_1(\vartheta)x].\end{aligned}$$

(Here the highest and lowest coefficients (with respect to x) are equal on the two sides.)

In final form, the system is

$$\left.\begin{aligned}(n-1)y_{n-1} - n\vartheta &= n\vartheta y'_{n-1} - \frac{y_1}{y'_1}\\ (n-2)y_{n-2} - \overline{n-1}\,y_{n-1} &= n\vartheta y'_{n-2} - \frac{y_1}{y'_1}y'_{n-1}\\ (n-3)y_{n-3} - (n-2)y_{n-2} &= n\vartheta y'_{n-3} - \frac{y_1}{y'_1}y'_{n-2}\\ \cdots\cdots&\cdots\cdots\\ y_1 - 2y_2 &= n\vartheta y'_1 - \frac{y_1}{y'_1}y'_2\end{aligned}\right\}$$

and the initial conditions are:

for $\vartheta = 2^{2n-1}\cos^{2n}\frac{\pi}{2n}$	for $\vartheta = 0$
$y_1 = t_1\cos^2\frac{\pi}{2n}$;	$y_1 = -t'_1$;
$y_2 = t_2\cos^4\frac{\pi}{2n}$;	$y_2 = -t'_2$;
$\cdots\cdots$	$\cdots\cdots$
$y_{n-1} = t_{n-1}\cos^{2n-2}\frac{\pi}{2n}$,	$y_{n-1} = -t'_{n-1}$;

where

$\sum_0^n t_i x^i = T_n(x)$.	$\sum_0^{n-1} t'_i x^i = T_{n-1}(x)$.

EXAMPLE 37. Find the analytic form of the $[3,3,0]$ polynomials. Since the polynomials $Z_3(x,\vartheta)$ are easily constructed algebraically, we shall do this and then verify that they satisfy (60).

For the algebraic construction it is convenient to take the deformation parameter to be the unique interior node of the polynomial; denote it by ρ. Then the distribution of the polynomial is $\bar{0},\overset{+}{\rho},\bar{1}$, where 0 and 1 are fixed, and ρ decreases on the interval $1/3 < \rho < 1/2$. As $\rho \to 1/3$, $Z_3(x,\rho) \to T_3(3x/4)$, and as $\rho \to 1/2$, $Z_3(x,\rho) \to -T_2(x)$.

The required polynomials are most easily found from the identity

$$(x-\rho)^2\phi_1(x) + x(1-x)\psi_1(x) \equiv 2,$$

(p. 38), where $\phi_1(x)$ and $\psi_1(x)$ are unknown linear functions.

Applying the Euclidean algorithm, we obtain after elementary calculations

$$\phi_1(x) = \frac{2(2\rho - 1)}{(1-\rho)^2\rho^2}x + \frac{2}{\rho^2}; \quad \psi_1(x) = \frac{2(2\rho-1)}{\rho^2(1-\rho)^2}x + \frac{2(2-3\rho)}{\rho(1-\rho)^2}.$$

Therefore we have

$$\vartheta = \frac{2(1-2\rho)}{(1-\rho)^2\rho^2}. \tag{61}$$

The required polynomial $Z_3(x,\rho)$ has the form

$$Z_3(x,\rho) = 1 - (x-\rho)^2\phi_1(x) = -1 + x(1-x)\psi_1(x),$$

i.e.

$$Z_3(x,\rho) = \frac{2(1-2\rho)}{\rho^2(1-\rho)^2}x^3 - 2\frac{1-3\rho^2}{\rho^2(1-\rho)^2}x^2 + 2\frac{2-3\rho}{\rho(1-\rho)^2}x - 1.$$

Thus the coefficients are differentiable with respect to ρ, and if we take account of (61) it is clear that the coefficients of $\xi_3(x,\vartheta)$ are differentiable with respect to ϑ. We still have to show that the coefficients satisfy both the differential equations (60) and the boundary conditions:

$$\text{for } \vartheta = \frac{27}{2}, \quad y_1 = -27, \quad y_2 = \frac{27}{2};$$

$$\text{for } \vartheta = 0, \quad y_1 = -8, \quad y_2 = 8$$

(since $T_3(x) = 32x^3 - 48x^2 + 18x - 1$ and $-T_2(x) = -8x^2 + 8x - 1$). In fact, in the present case the resolvent is $R_3(x,\rho) = x^3 - (1+\rho)x^2 + \rho x$. We must verify the identity

$$\frac{\partial \zeta_3(x,\vartheta)}{\partial \rho} \cdot \frac{\partial \rho}{\partial \vartheta} \equiv R(x,\rho);$$

but

$$\frac{\partial \vartheta}{\partial \rho} = -4\frac{3\rho^2 - 3\rho + 1}{\rho^3(1-\rho)^3},$$

whence it is easy to verify the two necessary identities for the coefficients. It is also easy to see that $\xi_3(x,\rho)$ satisfies

$$\xi_3\left(x, \frac{1}{2}\right) = -T_2(x) \quad \text{and} \quad \zeta_3\left(x, \frac{1}{3}\right) = T_3\left(\frac{3}{4}x\right).$$

Remark. The polynomial $\zeta_4(x,\vartheta)$, which is closely related to the polynomial of passport $[4,4,1(1)]$, will be constructed in Chapter 5, §2.

As we shall see in Part II, the polynomials just constructed are the Zolotarev polynomials.

§3. Some metric properties of $[n, n, 0]$ polynomials

THEOREM 42. *Each polynomial $Q_n(x)$ in the family of polynomials of passport $[n, n, 0]$ with distribution $(\overset{\pm}{\sigma}_i)_1^n$ is unique, and hence is the* principal *polynomial of the double nodal distribution $(\overset{\pm}{\sigma}_k, \overset{\mp}{\sigma}_{k+1})$, where σ_k, σ_{k+1}* are *adjacent interior nodes $(k = 2, 3, \cdots, n-2)$.*

In fact, let there be another polynomial $P_n(x)$ besides $Q_n(x)$ with the same double nodal distribution. Then by Theorem 15 there is an uncountable set of polynomials with the same distribution. They are all of the form

$$P_n(x) = Q_n(x) + \phi(x)(x - \sigma_k)^2(x - \sigma_{k+1})^2.$$

But $Q_n(x)$ has, besides σ_k and σ_{k+1}, an additional $n - 2$ points at which its values are alternately ± 1. Consequently the polynomial $\phi(x)$ must have at least $n - 3$ changes of sign on the interval $0 \leqq x \leqq 1$, whereas it is of degree $n - 4$. Since this is impossible, $Q_n(x)$ is unique.

COROLLARY. *A distribution $(\overset{\pm}{\sigma}_k, \overset{\mp}{\sigma}_{k+1})$ consisting of an arbitrary pair of adjacent nodes of the polynomial $Q_n(x)$ of passport $[n, n, 0]$ is a distribution of degree precisely n.*

In fact, if the chosen pair consists of interior nodes the assumption that the degree of the distribution is decreased implies multiplicity of the polynomial of a distribution of degree n, which is impossible. If, on the other hand, the nodes are a boundary pair, i.e. either $\sigma_k = 0$ or $\sigma_{k+1} = 1$, then the pair is close to a boundary pair of nodes of $T_{n-1}(x)$, and the degree of such a distribution is greater than $n - 1$ (Example 27 and Theorem 24).

THEOREM 43. *If $(\overset{\pm}{\sigma}_i)_1^n$ is the distribution of a polynomial $Q_n(x)$ of the family under consideration, and if $(\overset{\pm}{\rho}_1, \overset{\mp}{\rho}_2)$ is any double nodal distribution with the numbers $\rho_1 < \rho_2$ located in the interval (σ_k, σ_{k+1}) $(k = 2, 3 \cdots, n-2)$, then the distribution is of degree greater than n (i.e. it is fictitious for any segment $(\mu_i)_0^n$ constructed with this distribution).*

In fact, taking $(\overset{+}{\rho}_1, \overset{-}{\rho}_2)$, we choose the signs of $\pm Q_n(x)$ so that $\overset{+}{\sigma} \leqq \overset{+}{\rho}_1 < \overset{-}{\rho}_2 \leqq \overset{-}{\sigma}_{k+1}$ (equality in both places being excluded). Suppose that we have to select $+ Q_n(x)$. Let $P_{\underline{n}}(x)$ be an extremal polynomial of the distribution $(\overset{+}{\rho}_1, \overset{-}{\rho}_2)$. Then (p. 81) there is a transformation $\alpha x + \beta$ mapping the nodes (σ_k, σ_{k+1}) on (ρ_1, ρ_2), and consequently $P_n(\alpha x + \beta)$ and $Q_n(x, \Theta_0)$ would have a pair of nodes in common; this is impossible by a previous theorem.

If the pair (σ) is $(0, \sigma_2)$ $[$or $(\sigma_{n-1}, 1)]$ and we have $0 = \rho_1 < \tau_1 < \rho_2 < \tau_1'$, where τ_1 and τ_1' are respectively nodes of $T_n(x)$ and $T_{n-1}(x)$, the distribution $(\overset{+}{\rho}, \overset{-}{\rho})$ is precisely of degree n, with more than one principal polynomial if $\rho_2 > \tau_1/\cos^2(\pi/2n)$. In fact, both $T_n(\tau_1 x/\rho_2)$ and $Q_n(\alpha x)$ have the nodes $\rho_1 = 0$ and ρ_2 for suitable choice of α.

COROLLARY 1. *For the whole family $\{Q_n(x, \Theta)\}$ the nodes move to the right when Θ decreases from $\Theta'' = -\frac{1}{2}(n-1)$ to $\Theta' - \frac{1}{2}(n+1)$, and each node $\sigma_k(\Theta)$ runs through the interval from τ_k to τ_k', where τ_k and τ_k' are nodes of $T_n(x)$ and $T_{n-1}(x)$.*

COROLLARY 2. *No two polynomials of the family $\{Q_n(x, \Theta)\}$ can have common nodes except for the two fixed nodes 0 and 1, and each pair of adjacent interior nodes of one polynomial are incompatible with the corresponding pair for any other polynomial, i.e. cannot be mapped on them by a linear transformation.*

THEOREM 44. *When $n \geqq 3$, the interior nodes of the polynomials of passport $[n, n, 0]$ fill the interval $[0, 1]$. More precisely, the nodes of the polynomials $Q_n(x, \Theta)$ for $-\frac{1}{2}n \leqq \Theta \leqq -\frac{1}{2}(n-1)$ and the nodes of the polynomials $Q_n(1-x, \overline{\Theta})$ for $-\frac{1}{2}(n+1) \leqq \overline{\Theta} \leqq -\frac{1}{2}n$ together fill the interval $[0, 1]$ without duplication, except for the nodes $(\tau_k)_1^{n-1}$ and $(\tau_k')_1^{n-2}$, which occur doubly, i.e. for both families of polynomials.*

Before giving the proof in general we consider the special case $n = 3$. The distributions of $T_3(x)$ and $-T_2(x)$ are respectively $\overset{-}{0}, \overset{+}{\frac{1}{4}}, \overset{-}{\frac{3}{4}}, \overset{+}{1}$ and $\overset{-}{0}, \overset{+}{\frac{1}{2}}, \overset{-}{1}$. We vary the polynomial $Q_3(x, \Theta)$ of passport $[3, 3, 0]$ in the interval $-3/2 \leqq \Theta < -1$, with boundary values $Q_3(x, -1) = T_3(x)$; $Q_3(x, -3/2) = -T_2(x)$. Hence if we start from $T_3(x)$ the left interior node runs through the interval from $\frac{1}{4}$ to $\frac{1}{2}$, and the right-hand node runs from $\frac{3}{4}$ to 1. If we choose $Q_3(1-x, \overline{\Theta})$, then $-2 \leqq \overline{\Theta} \leqq -3/2$, with $Q_3(1-x, -3/2) = -T_2(x)$ and $Q_3(1-x, -2) = -T_3(x)$. If we start with $-T_3(x)$, the left-hand node runs through the interval from $\frac{1}{4}$ to 0, and the right-hand node runs from $\frac{3}{4}$ to $\frac{1}{2}$. Hence the conclusion of the theorem holds for $n = 3$.

We now turn to the general case.

By Theorems 43 and 44, the nodes $\sigma_2, \cdots, \sigma_{n-1}$ of $Q_n(x, \Theta)$ run monotonically through the intervals $[\tau_1, \tau_1']; [\tau_2, \tau_2']; \cdots; [\tau_{n-1}, 1]$ when $-\frac{1}{2}n \leqq \Theta \leqq -\frac{1}{2}(n-1)$, and the nodes $\overline{\sigma}_2 = 1 - \sigma_{n-1}, \cdots, \overline{\sigma}_{n-1} = 1 - \sigma_2$ of $Q_n(1-x, \overline{\Theta})$ run monotonically through $[0, \tau_1]; [\tau_1', \tau_2]; \cdots; [\tau_{n-2}', \tau_{n-1}]$ when $-\frac{1}{2}(n+1) \leqq \overline{\Theta} \leqq -\frac{1}{2}n$. This establishes the theorem.

The metric properties of distributions of passport $\lfloor n,n,0\rfloor$ established in Theorems 43 and 44 make it possible to define the degree of a given double nodal distribution.

THEOREM 45. *Let* $(\overset{+}{\rho}_1,\overset{-}{\rho}_2)$ *be an arbitrary double nodal distribution, with* $0<\rho_1<\rho_2<1$, *and find (in general, uniquely) by Theorem 23 two polynomials* $Q_n(x)$ *and* $Q_{n-1}(x)$ *of passports* $[n,n,0]$ *and* $[n-1,n-1,0]$ *such that* $Q_n(\rho_1)=+1$ *and* $Q_{n-1}(\rho_1)=+1$. *Let the nodes of* $Q_n(x)$ *be* $(\sigma_i^{(n)})$; *the nodes of* $Q_{n-1}(x)$, $(\sigma_j^{(n-1)})$; *and let* $\rho_1=\sigma_k^{(n)}=\sigma_j^{(n-1)}$. *Then a necessary and sufficient condition that the degree of the distribution* $\overset{+}{\rho}_1,\overset{-}{\rho}_2$ *is precisely* n *is that*

$$\sigma_{k+1}^{(n)}\leqq\rho_2<\sigma_{j+1}^{(n-1)}. \tag{62}$$

1. Let the degree of $(\overset{+}{\rho}_1,\overset{-}{\rho}_2)$ be precisely n, i.e. we have a polynomial $P_{\underline{n}}(x)$ with nodes $(\overset{+}{\rho}_1,\overset{-}{\rho}_2)$ and none of lower degree. We consider all the logically possible cases:

a) $\rho_2=\sigma_{k+1}^{(n)}$; then $P_n(x)\equiv Q_n(x)$, and by what has been proved, it is unique.

b) $\rho_2<\sigma_{k+1}^{(n)}$, i.e. the pair (ρ_1,ρ_2) is close to the pair $(\sigma_k^{(n)}=\rho_1,\sigma_{k+1}^{(n)})$. Then there is a transformation $\alpha x+\beta$ with fixed point ρ_1 which maps the second pair on the first. Then $P_n(\alpha x+\beta)\equiv Q_n(x)$ (by Theorem 37); but the inverse mapping yields $P_n(x)=Q_n(Ax+B)$, where $B<0$ and $A+B>1$. Since every $Q_n(x)$ is at least semiprimitive, the polynomial $Q_n(Ax+B)$ is not reduced. Thus case b) is impossible and we must have $\rho_2\geqq\sigma_{k+1}^{(n)}$.

c) Let $\rho_2\geqq\sigma_{j+1}^{(n-1)}$; if $\rho_2=\sigma_{n+1}^{(n-1)}$ then $Q_{n-1}(x)$ has the subdistribution $(\overset{+}{\rho}_1,\overset{-}{\rho}_2)$, which contradicts hypothesis 1. If on the other hand $\rho_2>\sigma_{j+1}^{(n-1)}$ then some transformation $ax+b$ yields $Q_{n-1}(ax+b)$, which contains the distribution $(\overset{+}{\rho}_1,\overset{-}{\rho}_2)$, and this is also impossible. Thus $\rho_2<\sigma_{j+1}^{(n-1)}$.

2. Let $\rho_1=\sigma_k^{(n)}=\sigma_j^{(n-1)}$ and $\sigma_{k+1}^{(n)}\leqq\rho_2<\sigma_{j+1}^{(n-1)}$. a) If $\rho_2=\sigma_{k+1}^{(n)}$ then $Q_n(x)$ is an extremal polynomial for the given double nodal distribution, and has the lowest possible degree; b) if $\rho_2>\sigma_{k+1}^{(n)}$ then a transformation $\alpha x+\beta$ yields $Q_n(\alpha x+\beta)$, a polynomial containing the nodes $(\overset{+}{\rho}_1,\overset{-}{\rho}_2)$. But by hypothesis $\rho_2<\sigma_{j+1}^{(n-1)}$ and hence the degree cannot be decreased, since the pair $(\overset{+}{\rho}_1,\overset{-}{\rho}_2)$ is close to $(\sigma_j^{(n-1)},\sigma_{j+1}^{(n-1)})$. Consequently the degree of the first pair exceeds $n-1$ (Theorem 23).

Thus condition (62) is both necessary and sufficient.

REMARK. It is clear that if a distribution $(\overset{\pm}{\sigma}_i)_1^s$ on $[0,1]$ is given, with $s>2$, then a necessary condition that the degree of the distribution is n is that there are not more than n pairs of adjacent nodes with equal signs. However, this condition is not sufficient. Even adding a single node with a permanence of sign to a distribution of degree n may increase the degree. We give an example.

Consider the distribution $(\overset{-}{0}, \overset{+}{\tfrac{1}{2}})$. Its degree is $n = 2$, since $-T_2(x)$ contains this distribution, and the degree clearly cannot be less. Consider the distribution $(\overset{-}{0}, \overset{+}{\tfrac{1}{2}}, \overset{+}{\rho})$ with three nodes, where $\rho = \frac{1}{2} + \epsilon < 1$. It is clear that this distribution has degree $n = 4$.

CHAPTER V

POLYNOMIALS OF PASSPORT $[n, n, 1]$

§1. Existence and nature of deformation of $[n, n, 1]$ polynomials

As we have already noted in Chapter III (pp. 70-71), the complete characteristic of polynomials of passport $[n, n, 1]$ must indicate the number of intervals of permanence; thus, if the distribution of such a polynomial is $(\overset{\pm}{\sigma}_i)_1^n$, where we have $\overset{+}{\sigma}_k$, $\overset{+}{\sigma}_{k+1}$, then $P = 1(k)$, and consequently the polynomials of this passport (if they exist) fall into $n-1$ classes corresponding to $k = 1, 2, \cdots, n-1$.

By the results of Chapter III, §1, the polynomials of a given class k form a family depending on a single deformation parameter; they are primitive, i.e. have the two fixed nodes $\sigma_1 = 0$ and $\sigma_n = 1$ and satisfy the boundary conditions (p. 81).

The existence of polynomials of this passport and of arbitrary class $(k = 1, 2, \cdots, n-1)$ can be established by constructing the corresponding segment functionals.

THEOREM 46. *The basis segment*

$$(\mu_i)_0^n = (-1)^{n-k+1}, (-1)^{n-k+1} \cdot \tau_k, (-1)^{n-k+1}\tau_k^2, \cdots, (-1)^{n-k+1}\tau_k^{n-1}, \qquad (63)$$
$$(-1)^{n-k+1}\tau_k^n + \Theta,$$

[where $\tau_k = \sin^2(k\pi/2n)$ $(k = 1, 2, \cdots, n-1)$ and $n \geqq 3$; $(-1)^{n-k}$ is the sign assigned to τ_k in the distribution of $+T_n(x)$] is served throughout the critical interval of Θ by polynomials $W_{n,1(k)}(x, \Theta)$ of passport $[n, n, 1(k)]$.

We first determine the values of Θ at the endpoints of the critical interval and at the focus of the segment (63). To do this, we decompose the segment $0_0, 0_1, \cdots, 0_{n-1}, \Theta$ in terms of the nodes $(\tau_i)_0^n$ of $T_n(x)$. We obtain

$$\delta_m = (-1)^{n-m} \frac{\Theta}{\prod_{i \neq m} |\tau_m - \tau_i|} \quad (m = 0, 1, \cdots, n),$$

and then we transpose the terms of the form $(\delta_k \tau_k^l)_{l=0}^n$ to the right-hand side and choose Θ so that $|\delta_k| = 1$. We thus obtain a decomposition of (63) in terms of the nodes $(\tau_i)_{i \neq k}$, where $\Theta = \prod_{j \neq k} |\tau_k - \tau_j| = \Theta''$.

In addition, it is clear that $\Theta^* = \Theta' = 0$, since when $\Theta = 0$ the segment has a quadratic as principal polynomial; $\Theta = 0$ is a point of multiplicity of extremal polynomials; $-T_n(x)$ must also occur among these extremal polynomials.

Since the segment (63) loses its weight at the node τ_k when $\Theta = \Theta''$, it follows that for $0 < \Theta < \Theta''$ the segment is served by a primitive polynomial with $s = n$ nodes and with signs having a single permanence $\overset{+}{\sigma}_k, \overset{+}{\sigma}_{k+1}$.

In the same way we can establish the existence of families of all classes of passport $[n, n, 1]$.

Corollary. *As $\Theta \to 0$, at least one of the two nodes $\sigma_k(\Theta)$ and $\sigma_{k+1}(\Theta)$ tends to τ_k, and $\delta_k(\Theta) \to 1$, whereas the other δ_m tend to zero. When both σ_k and σ_{k+1} tend to τ_k then $\delta_k + \delta_{k+1} \to 1$ and $\delta_m \to 0$ $(m \neq k, k+1)$. The latter case cannot occur for $k = 1$ or $k = n - 1$, since in these cases one node is fixed.*

Thus polynomials of passport $[n, n, 1(k)]$ exist for all k $(1, 2, \cdots, n-1)$. We let $\{W_{n,k}(x)\}$ denote the collection of polynomials of this passport, with the understanding that when the two nodes σ_k, σ_{k+1} have $+$ signs then the whole collection of polynomials of this passport is filled out by the polynomials $\{\pm W_{n,k}(x)\}$.

The collection $\{\pm W_{n,k}(1-x)\}$ contains all polynomials of passport $[n, n, 1(n-k)]$.

Theorem 46 is easily generalized by replacing τ_k by any number ρ satisfying $\overset{+}{\tau}_{k-1} < \rho < \overset{+}{\tau}_{k+1}$ (we use the signs $^{++}$ for definiteness).

Theorem 47. *In the interval $\Theta^* = 0 < \Theta < \Theta'' = \prod_{i \neq k} |\rho - \tau_i|$, the segment*

$$1, \rho, \rho^2, \cdots, \rho^{n-1}, \rho^n + \Theta \tag{64}$$

with $s = n$ nodes (σ_i) defines the subset of polynomials $\{W_n^{(\rho)}(x, \Theta)\} \subset \{W_{n,k}(x)\}$ for which $\overset{+}{\tau}_{k-1} \leqq \overset{+}{\sigma}_k < \rho < \overset{+}{\sigma}_{k+} \leqq \overset{+}{\tau}_{k+1}$, and no others.

The proof is a literal repetition of the proof of Theorem 46.

Here we also have the similar equation $\Theta = \prod_1^n |\rho - \sigma_i|$ for each $W_n^{(\rho)}(x, \Theta)$ with distribution $(\overset{\pm}{\sigma}_i)_1^n$.

We recall (Theorem 20) that as Θ decreases in the interval $(0, \Theta'')$ the leading coefficient of the extremal polynomial $q_n(\Theta)$ is a continuous and strictly decreasing function, and as $\Theta \to 0+$, $W_n^{(\rho)}(x, \Theta)$ approaches a limit with leading coefficient $q_n(0) > 0$; and at least one of the nodes $\sigma_k(\Theta)$ and $\sigma_{k+1}(\Theta)$ tends to ρ as $\Theta \to 0+$. Therefore $\Theta = \phi(q_n)$ is also a continuous monotonic function and $\{W_n^{(\rho)}(x,\Theta)\} \equiv \{W_n^{(\rho)}(x,q)\}$.

Let two segments

(65) $\quad 1, \rho_1, \rho_1^2, \cdots, \rho_1^{n-1}, \rho_1^n + \Theta_1$ and $1, \rho_2, \rho_2^2, \cdots, \rho_2^{n-1}, \rho_2^n + \Theta_2$

be given, where $\tau_{k-1} \leqq \rho_1, \rho_2 \leqq \tau_{k+1}$.

By Theorem 47, each of them defines, in the respective intervals

$$0 \leqq \Theta_1 < \prod_{i \neq k} |\rho_1 - \tau_i| = \Theta_1'' \quad \text{and} \quad 0 \leqq \Theta_2 < \prod_{i \neq k} |\rho_2 - \tau_i| = \Theta_2'',$$

the subsets

$$\{W_n^{(\rho_1)}(x)\} \subset \{W_{n,k}(x)\} \quad \text{and} \quad \{W_n^{(\rho_2)}(x)\} \subset \{W_{n,k}(x)\},$$

and the leading coefficients in these subsets run through the corresponding intervals $(A, 2^{2n-1})$ and $(B, 2^{2n-1})$; for definiteness, we suppose that $A < B$.

Theorem 48. *The polynomials with the same leading coefficients in the sets* $\{W_n^{(\rho_1)}(x)\}$ *and* $\{W_n^{(\rho_2)}(x)\}$ *are identically equal.*

Suppose that for a certain q in $(B, 2^{2n-1})$ we have $W_n^{(\rho_1)}(x,q) \not\equiv W_n^{(\rho_2)}(x,q)$ and put $\Theta_1 = \phi_1(q)$; $\Theta_2 = \phi_2(q)$.

Then by the theorem on continuous deformation we have an interval near Θ_1 and Θ_2 where $W_n^{(\rho_1)}(x,\Theta_1) \not\equiv W_n^{(\rho_2)}(x,\Theta_2)$, and this interval is open on the right. Let $\Theta_1^{(0)}$ and $\Theta_2^{(0)}$ be corresponding points where $W_n^{(\rho_1)}(x,\Theta_1^{(0)}) \equiv W_n^{(\rho_2)}(x,\Theta_2^{(0)})$, or, what is the same thing, $W_n^{(\rho_1)}(x,q^{(0)}) \equiv W_n^{(\rho_2)}(x,q^{(0)})$, but $W_n^{(\rho_1)}(x,q^{(0)} - \Delta q) \not\equiv W_n^{(\rho_2)}(x,q^{(0)} - \Delta q)$ for arbitrarily small $\Delta q > 0$. By hypothesis the polynomial $W_n^{(\rho_1)}(x,q^{(0)} - \Delta q)$ serves the first segment (65) for $\Theta_1 = \phi_1(q^{(0)} - \Delta q)$; but $W_n^{(\rho_1)}(x,q^{(0)} - \Delta q)$ serves the second segment (65) at the point $q^{(0)} - \Delta q$, since for sufficiently small Δq its nodes σ_k and σ_{k+1} satisfy $\sigma_k < \rho_1$, $\rho_2 < \sigma_{k+1}$, and then necessarily

$$W_n^{(\rho_2)}(x,q^{(0)} - \Delta q) \equiv W_n^{(\rho_1)}(x,q^{(0)} - \Delta q).$$

Thus equality of the leading coefficients in $\{W_n^{(\rho_1)}(x)\}$ and $\{W_n^{(\rho_2)}(x)\}$ implies identity of the polynomials, as was to be proved.

Corollary 1. *The set* $\{W_{n,k}(x)\}$ *can be represented as* $W_{n,k}(x,q_n)$, *i.e. it is a one-parameter set, where the leading coefficient can be taken as the continuous monotonic parameter, and the whole domain of* $q_n = q$ *is obtained for the* ρ *that yields the smallest left-hand boundary of the corresponding values of* Θ, *or the smallest possible* $q = q_{\min} = q_0$.

Corollary 2. *Each* $W_n(x) \subset \{W_{n,1(k)}(x)\}$ *is determined by specifying either its left-hand node* $\sigma_k > 0$ *or its right-hand node* $\sigma_{k+1} < 1$.

In fact, put $\rho = \sigma_k$; the segment $1, \sigma_k, \cdots, \sigma_k^{n-1}, \sigma_k^n + \Theta$ defines $W_n(x) = W_n(x, \sigma_k)$ uniquely as $\Theta \to 0$. Therefore $\sigma_k = \Psi(q)$ is monotonic and continuous, and consequently

$$W_{n,k}(x, q) \equiv W_{n,k}(x, \sigma_k).$$

Since we have $\sigma_k = \tau_{k-1}$ for $q = q^{2n-1}$, and we must have $\sigma_k > \tau_{k-1}$ when q is decreased, it follows that Ψ is a decreasing function of q.

Similarly σ_{k+1} is an increasing function of q_n.

Corollary 3. *There is a single point ρ_k^* with $\tau_{k-1} < \rho_k^* < \tau_{k+1}$ such that $\sigma_k < \rho_k^* < \sigma_{k+1}$ and*

$$\lim_{q \to q_0} \sigma_k = \lim_{q \to q_0} \sigma_{k+1} = \rho_k^*$$

for all polynomials $\{ W_{n,k}(x) \}$.

In fact, as q_n decreases, the nodes $\sigma_k(q)$ and $\sigma_{k+1}(q)$ move toward each other; if

$$\sigma_k(q_0) = \lim_{q \to q_0} \sigma_k(q) \quad \text{and} \quad \sigma_{k+1}(q_0) = \lim_{q \to q_0} \sigma_{k+1}(q)$$

with $\sigma_k(q_0) < \sigma_{k+1}(q_0)$, then by taking $\sigma_k(q_0) < \rho < \sigma_{k+1}(q_0)$ we find that as $\Theta \to 0+$ the segment $1, \rho, \cdots, \rho^n + \Theta$ determines a polynomial $W_n(x) \subset \{ W_{n,k}(x) \}$ with a left-hand or right-hand node ρ, which is impossible by hypothesis. Thus $\sigma_k(q)$ and $\sigma_{k+1}(q)$ have the common limit ρ_k^*.

Corollary 4. *The basis segment*

$$1, \rho_k^*, \rho_k^{*2}, \cdots, \rho_k^{*n-1}, \rho_k^{*n} + \Theta, \tag{66}$$

with $0 \leqq \Theta < \Theta''$ is, when $k = 2, 3, \cdots, n-2$, a determining segment for the polynomials $W_{n,k}(x, \Theta) \cdot (-1)^{n-k+1}$, and only for these.

In fact, the point ρ_k^*, being located between the two nodes σ_k, σ_{k+1} of every polynomial $W_{n,k}(x)$, defines the polynomials of passport $[n, n, 1(k)]$ by the functional (66), and obviously only these polynomials.

Corollary 5. *The number ρ_k^* for a determining segment can be found from the condition that the length of the critical interval for Θ'' is as large as possible, i.e. $|\prod_{i \neq k} (\rho - \tau_i)| = \max(\rho)$ for $\tau_{k-1} \leqq \rho \leqq \tau_{k+1}$. As $\Theta \to 0+$, (66) defines a polynomial of passport $[n, n-1, 0]$.*

Remark. For the passports $[n, n, 1(1)]$ and $[n, n, 1(n-1)]$ there are no determining basis segments (in the sense of complete determination).

The existence of polynomials of these two classes was established in

Theorem 46. On the basis of the preceding discussion we can state that the segment $1, \rho, \rho^2, \cdots, \rho^{n-1}, \rho^n + \Theta$ with $\tau_{k-1} < \rho < 1$ defines the family of passport $[n, n, 1(k-1)]$, and as $\rho \to 1$ this family becomes complete.

§2. Methods of constructing $[n, n, 1]$ polynomials in the simplest cases

We shall give some examples of direct calculation of the polynomials $W_{n,k}(x)$ for small n.

We note that the smallest possible degree for such polynomials is $n = 3$. In our notation they are $W_{3,1}$ and $W_{3,2}$. By analogy with the polynomials $W_{n,k}$ $(n \geqq 3)$ we may take $W_{2,1}(x, \vartheta)$ to be the polynomial $1 - \vartheta x(1-x)$ with $0 < \vartheta < 8$, but it is not the principal polynomial for any non-absolutely-monotone segment $(\mu_i)_0^2$. It will be needed in Example 43.

EXAMPLE 38. The polynomial $W_{3,2}$ is easily constructed if we take the parameter ρ to be the single interior node of the polynomial. Then

$$W_{3,2} = W_{3,2}(x, \rho) = 1 - \vartheta(x-\rho)^2(1-x), \ \tfrac{1}{4} < \rho < 1.$$

Since $W_{3,2}(0, \rho) = -1$, we have $\vartheta = 2/\rho^2$, and the second root of the derivative is $\lambda = (\rho + 2)/3$. Thus

$$W_{3,2}(x, \rho) = \frac{2}{\rho^2} x^3 - \frac{2(2\rho+1)}{\rho^2} x^2 + \frac{2(\rho+2)}{\rho} x - 1.$$

The resolvent is $R_{3,2}(x, \rho) = x^3 - (1+\rho)x^2 + \rho x$.

For $\rho = \lambda = 1$ (i.e. for $\Theta = 0$) we obtain the polynomial

$$\overline{W}_{3,2}(x) = 2x^3 - 6x^2 + 6x - 1 = 1 - 2(1-x)^3.$$

As $\rho \to \frac{1}{4}$ we have $W_{3,2}(x, \rho) \to + T_3(x)$.

Clearly $W_{3,1}(x, \rho) = - W_{3,2}(1-x, \rho)$, and the polynomials $\pm W_{3,2}(x, \rho)$, $\pm W_{3,2}(1-x, \rho)$ exhaust the polynomials of passport $[3, 3, 1]$. If we take the leading coefficient ϑ as parameter, and denote the polynomial by $V_{3,2}(x, \vartheta)$, we have

$$V_{3,2}(x, \vartheta) = \vartheta x^3 - (2\sqrt{2\vartheta} + \vartheta)x^2 + 2(1 + \sqrt{2\vartheta})x - 1.$$

EXAMPLE 39. We construct the polynomials of passport $[4, 4, 1]$.

To construct $W_{4,1}$, we let ρ denote the interior node at which $W_{4,1} = +1$, and let γ denote the second interior node, $\rho < \gamma$. Then $W_{4,1}(x, \rho) = 1 - \vartheta x(1-x)(x-\rho)^2$ and for $x = \gamma$ we have

$$1 - \vartheta\gamma(1-\gamma)(\gamma-\rho)^2 = -1.$$

Since $\partial W_{4,1}/\partial x = 0$ at γ, we have $4\gamma^2 - (2\rho+3)\gamma + \rho = 0$, and hence

$$\rho = \frac{\gamma(4\gamma - 3)}{2\gamma - 1};$$

consequently ϑ and ρ are rational functions of γ; we use γ as deformation parameter.

We have

$$W_{4,1}(x, \rho) = \vartheta x^4 - \vartheta(1 + 2\rho)x^3 + \vartheta\rho(2 + \rho)x^2 - \vartheta\rho^2 x + 1,$$

where

$$\vartheta = \frac{(2\gamma - 1)^2}{2\gamma^3(1 - \gamma)^3};$$

as $\rho \to 0$ we have $\gamma \to \frac{3}{4}$, and we obtain the following expression for the limiting polynomial:

$$\overline{W}_{4,1}(x) = \frac{512}{27}x^4 - \frac{512}{27}x^3 + 1 = 1 - \frac{512}{27}x^3(1 - x)$$

As $\rho \to \frac{1}{2}$, i.e. as $\vartheta \to 128$, we have $W_{4,1}(x, \rho) \to T_4(x)$.

Clearly $W_{4,3}(x, \rho) = W_{4,1}(1 - x, \rho)$, and similarly $\overline{W}_{4,3}(x) = \overline{W}_{4,1}(1 - x)$.

It remains to find $W_{4,2}(x)$, which by symmetry has the form $W_{4,2}(x) = -1 + \vartheta(x - \rho)^2(x - 1 + \rho)^2$, and since $W_{4,2}(0) = 1$, we have

$$\vartheta = \frac{2}{\rho^2(1 - \rho)^2}$$

(here ρ is one of the interior nodes).

The third root of $\partial W / \partial x$ is fixed: $\lambda = \frac{1}{2}$.

Thus with ϑ as parameter we have

$$V_{4,2}(x, \vartheta) = \vartheta x^4 - 2\vartheta x^3 + (\vartheta + 2\sqrt{2\vartheta})x^2 - 2\sqrt{2\vartheta}x + 1.$$

Here $32 < \vartheta < 128$; at the boundaries we obtain

$$V_{4,2}(x, 128) = T_4(x);$$
$$V_{4,2}(x, 32) = \overline{V}_{4,2}(x),$$

where

$$\overline{V}_{4,2}(x) = 32x^4 - 64x^3 + 48x^2 - 16x + 1 = -1 + 32(x - \tfrac{1}{2})^4.$$

EXAMPLE 40. We find the limiting polynomial in the family $V_{5,4}(x, \vartheta)$, where ϑ is the leading coefficient.

Denote the required polynomial by $\overline{V}_{5,4}(x)$. Then by the nature of the deformation $\overline{V}_{5,4}(x) = 1 - \vartheta_0(x-\rho_0)^2(1-x)^3$, where ϑ_0 and ρ_0 are numerical constants that can be calculated. In addition, $\overline{V}_{5,4}(0) = -1$ and $\overline{V}_{5,4}(\lambda_0) = -1$, where $\lambda_0 = (3\rho_0+2)/5$ is the second root of $\partial \overline{V}_{5,4}/\partial x$. Then $\vartheta_0 = 2/\rho_0^2$ and ρ_0 is defined by the equation $3125\rho_0^2 - 108(1-\rho_0)^5 = 0$; this equation has a unique root in $0.1 < \rho_0 < 0.2$.

EXAMPLE 41. The polynomial $\overline{V}_{6,1}(x) = 1 - \vartheta_0 x^3(1-x)(x-\rho_0)^2$ where ϑ_0 and ρ_0 are numerical constants. If λ_1 denotes one of the nodes at which $\overline{V}_{6,1}(x) = -1$, the other one is $\lambda_2 = \frac{1}{2}\rho_0/\lambda_1$. Then the equations

$$\rho_0 = \frac{6\lambda_1^2 - 5\lambda_1}{4\lambda_1 - 3}; \quad \vartheta_0 = \frac{(4\lambda_1-3)^2}{2\lambda_1^5(1-\lambda_1)^3} \quad \text{and} \quad \overline{V}_{6,1}\left(\frac{\rho_0}{2\lambda_1}\right) = -1$$

make it possible to find λ_0, ϑ_0 and ρ_0.

EXAMPLE 42. We find the relation between the polynomials $V_{4,1}(x,\vartheta)$ and $\zeta_4(x,\vartheta')$, where ϑ and ϑ' are the leading coefficients of the polynomials and are taken as deformation parameters. We can pass from one family to the other by using a linear mapping that preserves the coordinate axes and changes the scales (as is obvious graphically) (p. 98).

In fact, if λ is the smallest root of $\partial V_{4,1}(x,0)/\partial x$ and $x=\alpha$ is the smallest simple root of $V_{4,1}(x,\vartheta) = V_{4,1}(\lambda,\vartheta)$, we have

$$\zeta_4(x,\vartheta') = \frac{2}{1 - V_{4,1}(\lambda,\vartheta)} V_{4,1}(\alpha x,\vartheta) - \frac{1 + V_{4,1}(\lambda,\vartheta)}{1 - V_{4,1}(\lambda,\vartheta)}. \tag{67}$$

Hence the relation between ϑ' and ϑ is

$$\vartheta' = \frac{2\alpha^4 \cdot \vartheta}{1 - V_{4,1}(\lambda,\vartheta)},$$

where λ is the smaller root of

$$4x^2 - (2\rho+3)x + \rho = 0$$

(Example 39).

It is easily verified that the polynomial (67) takes the value $+1$ at $x=0$ and $x = \rho/\alpha < 1$, and the value -1 at $x=1$ and $x = \lambda/\alpha < \rho/\alpha$; but λ/α and ρ/α are the roots of $\partial V(\alpha x,\vartheta)/\partial x$, and hence are the only ones on $[0,1]$.

Hence the maximum modulus of the polynomial (67) on $[0,1]$ is unity, and its passport is $[4,4,0]$.

In questions involving the analytic form of $W_{n,k}(x,\Theta)$ $(k = 1,2,\cdots,n-1)$ it is convenient to replace the parameter Θ that we chose as the parameter of the deformation by the leading coefficient of the polynomial $W_{n,k}$, which is continuous and increases with Θ in the interval $0 < \Theta < \Theta''$

(Theorem 20). After this change we denote the family of polynomials by $V_{n,k}(x, \vartheta)$ or $V_{[n,n,1(k)]}(x)$.

In some cases the polynomials $V_{n,k}(x, \vartheta)$ can be expressed in an elementary way in terms of Čebyšev polynomials and V polynomials of lower degree.

Theorem 49. *The following identity holds for polynomials of passport $[N, N, 1(K)]$ when N and K have the common divisor m:*

$$V_{[N,N,1(K)]}(x, \vartheta') = T_m\left[\frac{1 + V_{[n,n,1(k)]}(x, \vartheta)}{2}\right], \tag{68}$$

where $N = m \cdot n$ and $K = m \cdot k$. The leading coefficients ϑ' and ϑ of the respective polynomials are related by

$$\vartheta' = 2^{m-1} \cdot \vartheta^m.$$

Let $V_{[n,n,1(k)]}(x, \vartheta)$ have the distribution $(\overset{\pm}{\sigma}_i)_1^n$. Then (σ_k, σ_{k+1}) is (the unique) interval of permanence. Suppose for definiteness that we have $\overset{+}{\sigma}_k$ and $\overset{+}{\sigma}_{k+1}$. The polynomial

$$Y_n(x, \vartheta) = \frac{1 + V_{[n,n,1(k)]}(x, \vartheta)}{2}$$

satisfies $0 \leqq Y_n(x, \vartheta) \leqq 1$; for $0 \leqq x \leqq 1$,

$$Y_n(\overset{+}{\sigma}_i, \vartheta) = +1; \quad Y_n(\overset{-}{\sigma}_i, \vartheta) = 0.$$

Let $\lambda = \lambda(\vartheta)$ be the unique root of $\partial V/\partial x$ in the interval (σ_k, σ_{k+1}) and let $\tau_1 = \sin^2(\pi/2m)$ be the first node of $T_m(x)$.

Then $T_m[Y_n(x, \vartheta)]$ is a polynomial of degree N which has $N = n \cdot m$ nodes for $0 < Y_n(\lambda, \vartheta) < \tau_1$. These nodes form the distribution $(\overset{\pm}{\rho}_i)_1^N$, which contains just one permanence of sign, the interval of permanence being (ρ_{mk}, ρ_{mk+1}).

Consequently the passport of $T_m[Y_n]$ is $[N, N, 1(K)]$, and (68) is established.

[If the interval is $(\overset{-}{\sigma}_k, \overset{-}{\sigma}_{k+1})$ then we consider $\tau_{m-1} < Y_n(\lambda, \vartheta) < 1$, where $\tau_{m-1} = \cos^2(\pi/2m)$.]

The relation between ϑ' and ϑ is obtained immediately by equating leading coefficients of the two sides of (68).

Corollary. *For $Y_n(\lambda, \vartheta) = \tau_1$ the two nodes coincide: $\rho_{mk} = \rho_{mk+1}$, and $V(x, \vartheta')$ becomes $\overline{V}(x)$, of passport $[N, N-1, 0]$. Formula (68) holds for $\vartheta_0 \leqq \vartheta \leqq 2^{2n-1}$, where ϑ_0 is the unique double root of*

$$Y_n[\lambda(\vartheta), \vartheta] = \tau_1.$$

EXAMPLE 43. The polynomials $V_{6,3}(x, \vartheta')$ and $V_{6,4}(x, \vartheta')$ can be constructed by Theorem 49, since $V_{2,1}(x, \vartheta) = 1 - \vartheta x(1-x)$ and $V_{3,2}(x, \vartheta)$ are known; in fact,

$$V_{6,3}(x, \vartheta') = T_3\left[\frac{1 + V_{2,1}(x, \vartheta)}{2}\right] = T_3\left[1 - \frac{\vartheta}{2}x(1-x)\right],$$

which yields $\vartheta' = 4\vartheta^3$ and $6 < \vartheta < 8$. For $\vartheta = 8$ we have

$$T_3\left[\frac{1 + T_2(x)}{2}\right] = T_6(x);$$

and for $\vartheta = 6$

$$T_3[1 - 3x(1-x)] = \overline{V}_{6,3}(x) = 864x^6 - + \cdots$$
$$= 1 - 864x(1-x)\left(x - \frac{1}{2}\right)^4;$$

$$V_{6,4}(x, \vartheta') = T_2\left[\frac{1 + V_{3,2}(x, \vartheta)}{2}\right],$$

where $\vartheta_0 < \vartheta < 32$ and ϑ_0 is the root of

$$\frac{1 + V_{3,2}(\lambda, \vartheta)}{2} = \frac{1}{2},$$

i.e. of $V_{3,2}(\lambda, \vartheta) = 0$. Here

$$\lambda(\vartheta) = \frac{1}{3}\left(\sqrt{\frac{2}{\vartheta}} + 2\right)$$

(Example 38) and $\vartheta' = 4\vartheta$. For $\vartheta = 32$ we have

$$T_2\left[\frac{1 + T_3(x)}{2}\right] = T_6(x);$$

for $\vartheta = \vartheta_0$ we obtain $\overline{V}_{6,4}(x)$; but for $V_{6,4}(x) = V_{6,4}(x, \vartheta_0')$ we have $\vartheta_0' = 51200/81$.

Consequently $\vartheta_0 = \frac{1}{4}\vartheta_0' = 12800/81$. $[\overline{V}_{6,4}(x) = \overline{V}_{6,2}(1-x).]$

EXAMPLE 44. The polynomials $V_{2n,n}(x, \vartheta')$ are equal to

$$T_n\left[\frac{1 + V_{2,1}(x, \vartheta)}{2}\right].$$

Consequently

$$V_{2n,n}(x, \vartheta) = T_n\left[1 - \frac{\vartheta}{2}x(1-x)\right],$$

where $\vartheta' = 2^{n-1}\vartheta^n$ and $8\cos^2(\pi/2n) < \vartheta < 8$.

For $\vartheta = 8$ we have $T_{2n}(x) = T_n[1 - 4x(1-x)]$; for $\vartheta = 8\cos^2(\pi/2n)$ we obtain

$$\overline{V}_{2n,n}(x) = T_n\left[1 - 4\cos^2\frac{\pi}{2n}\, x(1-x)\right].$$

It is easily shown that in these polynomials $\lambda(\vartheta')$ is the unique root of $\partial V_{2n,n}/\partial x$ lying in the interval of permanence, not equal to a node of $V_{2n,n}$; it is fixed and equal to $\frac{1}{2}$. This is not true in other cases, as can be seen immediately for the polynomials $V_{6,4}(x, \vartheta')$ and in general from (68).

In fact, the interval (ρ_{mk}, ρ_{mk+1}) of permanence of $V_{[N,N,1(k)]}(x, \vartheta')$ lies inside the interval (σ_k, σ_{k+1}) of permanence of $V_{[n,n,1(k)]}(x, \vartheta)$. We have

$$\frac{\partial V_N}{\partial x} = T'_m[Y_n] \cdot \frac{\partial Y_n}{\partial x} = T'_m[Y_n] \cdot \frac{1}{2}\frac{\partial V_n}{\partial x},$$

and therefore $\lambda(\vartheta)$ is the unique root of $\partial V_N/\partial x$ in the interval of permanence.

§3. General analytic construction of $[n, n, 1]$ polynomials

THEOREM 50. *If we take, for definiteness, k so that $T_n(\tau_k) = +1$, i.e. $k \equiv n \pmod 2$, then each of the n segments* (p. 106)

$$(69)\quad \begin{aligned} (\mu_i)_0^n &= -1, -\rho_k^*, -\rho_k^{*2}, \cdots, -\rho_k^{*n-1}, -\rho_k^{*n} + \Theta \\ (\mu_i^{(1)})_0^n &= -1, -\rho_k^*, -\rho_k^{*2}, \cdots, -\rho_k^{*n-1} - \Theta^{(1)}, -\rho_k^{*n} \\ &\cdots\cdots\cdots\cdots\cdots\cdots\cdots\cdots \\ (\mu_i^{(n-1)})_0^n &= -1, -\rho_k^* + (-1)^{n-1}\Theta^{(n-1)}, -\rho_k^{*2}, \cdots, -\rho_k^{*n-1}, -\rho_k^{*n}, \end{aligned}$$

(where $\Theta, \Theta^{(1)}, \cdots, \Theta^{(n-1)}$ are variables not in the corresponding critical intervals) is a determining segment for the family of polynomials $W_{n,k}(x, \Theta)$ determined by the first segment. If we put $W_{n,k}(x, \Theta) = \sum_{i=0}^n \beta_{i,k}(\Theta)x^i$, the relation between the parameters $\Theta^{(l)}$ $(l = 1, 2, \cdots, n-1)$ and the fundamental parameter Θ is

$$(70)\quad \Theta^{(l)} = (-1)^l \cdot \frac{\partial W_{n,k}(\rho_k^*, \Theta)/\partial\Theta}{\beta'_{n-l,k}(\Theta)}.$$

(The differentiability of the coefficients will be established in Theorem 51.)

We can find the endpoints of the critical interval for each segment exactly as for the segment $(\mu_i)_0^n$ in Theorem 46; i.e. we decompose the segment $0_0, \cdots, 0_{l-1}, x, 0_{l+1}, \cdots, 0_n$ in terms of the nodes $(\tau_i)_0^n$ $(i \neq k)$ with the additional node ρ_k^*.

We put

$$\delta_m = (-1)^{m+l} \frac{S_{n-l}^{(m)} \cdot x}{|\tau_m - \rho_k^*| \prod_{i \neq m_1 k} |\tau_m - \tau_i|} \quad (m \neq k)$$

(where the values of the $S_{n-l}^{(m)}$ are given on p. 47). Then

$$\delta_k = (-1)^{k+l} \frac{S_{n-l}^{(k)} \cdot x}{\prod |\rho_k^* - \tau_i|} \quad (i \neq k),$$

and in particular $\delta_k = +1$, which yields

$$x = (-1)^{l+k} \cdot \Theta^{(l)}, \quad \text{where} \quad \Theta^{(l)} = \frac{\prod_{i \neq k} |\rho_k^* - \tau_i|}{S_{n-l}^{(k)}} > 0.$$

Transposing the terms $(\rho_k^{*i})_{i=0}^n$ to the right-hand side, we obtain the decomposition of the segment

$$-1, -\rho_k^*, \cdots, -\rho_k^{*l} + (-1)^{l+k}\Theta^{(l)}, \cdots, -\rho_k^{*n}.$$

In this decomposition the (left-hand) node ρ_k^* loses its weight, and consequently

$$\Theta^{(l)''} = \frac{\prod_{i \neq k} |\rho_k^* - \tau_i|}{S_{n-l}^{(k)}}$$

(the right-hand endpoint of the critical interval).

The focus of the segment is clearly $\Theta^{(l)} = 0$. Thus

$$0 < \Theta^{(l)} < \frac{\prod_{i \neq k} |\rho_k^* - \tau_i|}{S_{n-l}^{(k)}}$$

is the domain of the parameters.

According to Theorem 47, the passport of each of the segments under consideration is $[n, n, 1(k)]$ in the right-hand part of the critical interval. Every segment similar to the first one $(k \neq 1,\ k \neq n-1)$ determines (up to sign) the polynomials of this passport for which the deformation begins with $+T_n(x)$ at $\Theta = \Theta''$.

Thus the segments (69) form a system of equivalent determining segments.

If $k + 1 \equiv n \pmod 2$, the system (69) is a system of equivalent segments for the polynomials $-W_{n,k}(x, \Theta)$ determined by the first segment.

The relation between the parameters can be found from

$$W_{n,k}(\mu^{(l)}, \Theta) = \max(\Theta);$$

i.e.

$$-W_{n,k}(\rho_k^*, \Theta) + (-1)^l \beta_{n-l,k}(\Theta) \cdot \Theta^{(l)} = \max(\Theta).$$

Hence we obtain formula (70).

A similar system (69) of equivalent segments is appropriate for $W_{n,1(1)}(x)$ or $W_{n,1(n-1)}(x)$ for $0 < \rho < \tau_2$ and $\tau_{n-2} < \rho < 1$, except that the resulting family W_n is not complete in the interval $0 < \Theta < \Theta''$.

We are going to replace the parameter Θ by ϑ, which (p. 107) is the leading coefficient of the polynomial $W_{n,k}(x, \Theta) = V_{n,k}(x, \vartheta)$.

Theorem 51. *If we put* $V_{n,k}(x, \vartheta) = \sum_0^n y_i(\vartheta) x^i$, *and*

$$R_{n,k}(x, \vartheta) = \prod_1^n (x - \sigma_i) = \sum_1^{n=1} \alpha_i(\vartheta) x^i + x^n$$

is the resolvent of $V_{n,k}$, *then* $y_i(\vartheta)$ *is differentiable with respect to* ϑ:

$$y_i'(\vartheta) = \alpha_i(\vartheta) \quad \textit{and} \quad \frac{\partial V_{n,k}(x, \vartheta)}{\partial \vartheta} = R_{n,k}(x, \vartheta). \tag{71}$$

Using the segment (69), we have by the properties of the resolvent

$$-R_{n,k}(\rho_k^*, \vartheta) + \Theta = 0; \quad -R_{n,k}(\rho_k^*, \vartheta) - \alpha_{n-1}(\vartheta) \cdot \Theta^{(1)} = 0,$$

etc.

On the other hand, the dependence of $\Theta, \Theta^{(1)}, \cdots, \Theta^{(n-1)}$ on ϑ can be found from the conditions $-V_{n,k}(\rho_k^*, \vartheta) + \Theta \cdot \vartheta = \max(\vartheta)$, i.e. $V_{n,k}(\rho_k^*, \vartheta) - y_{n-1}(\vartheta) \cdot \Theta^{(1)} = \max(\vartheta)$, etc. Then we have

$$\Theta = \frac{\partial V_{n,k}(\rho_k^*, \vartheta)}{\partial \vartheta};$$

$$\Theta^{(1)} = -\frac{\partial V_{n,k}(\rho_k^*, \vartheta)/\partial \vartheta}{y_{n-1}'(\vartheta)}; \cdots; \Theta^{(n-1)} = (-1)^{n-1} \frac{\partial V_{n,k}(\rho_k^*, \vartheta)/\partial \vartheta}{y_i'(\vartheta)}.$$

Hence

$$\frac{\partial V_{n,k}(\rho_k^*, \vartheta)}{\partial \vartheta} = R_{n,k}(\rho_k^*, \vartheta) \quad \text{and} \quad y_{n-1}'(\vartheta) = \alpha_{n-1}(\vartheta), \cdots, y_1'(\vartheta) = \alpha_1(\vartheta).$$

Since the extreme coefficients of $V_{n,k}$ and $R_{n,k}$ satisfy $\alpha_n(\vartheta) = 1 = (\vartheta)'$ and $\alpha_0(\vartheta) = 0 = [(-1)^n]'$, we obtain formula (71) in general.

Thus the differential equation (71) connecting an extremal polynomial and its resolvent is, for every subclass of the polynomials of passport $[n, n, 1(k)]$, the same as for the family $\zeta_n(x, \vartheta)$ of passport $[n, n, 0]$, according to Theorem 41.

Similarly the system of differential equations for $V_{n,k}(x, \vartheta)$ has the same form as the system (60), differing only in the supplementary conditions.

In fact, here also we have the following relation between $R_{n,k}(x, \vartheta)$ and $V_{n,k}(x, \vartheta)$: if λ denotes a root of $\partial V_{n,k}/\partial x$ that is not a node of $V_{n,k}(x, \vartheta)$, then

$$R_{n,k}(x, \vartheta) = \frac{\partial V_{n,k}(x, \vartheta)}{\partial x} \cdot \frac{x(x-1)}{n\vartheta(x-\lambda)}.$$

Here λ can be expressed, just as in Chapter 4, §2, by

(72) $$\lambda = \frac{y_1(\vartheta)}{n\vartheta \cdot \alpha_1(\vartheta)}, \text{ but here } \tau_{k-1} < \lambda < \tau_{k+1}.$$

Hence we obtain the system of differential equations

(73) $$x(x-1)\frac{\partial}{\partial x} V_{n,k}(x, \vartheta) = \left(n\vartheta x - \frac{y_1(\vartheta)}{y_1'(\vartheta)}\right) \cdot \frac{\partial}{\partial \vartheta} V_{n,k}(x, \vartheta),$$

which are identical in form with (60).

The condition at the right-hand end of the critical interval is this: when $\vartheta = 2^{2n-1}$ we have $V_{n,k}(x, \vartheta) = T_n(x)$, i.e.

$$y_i(2^{2n-1}) = t_i \quad \text{for} \quad i = 1, 2, \cdots, n-1.$$

Since this is not required in the system of differential equations for $\zeta_n(x, \vartheta)$, these conditions exclude the polynomials $\zeta_n(x, \vartheta)$ from the integrals of the system; for the polynomials $V_{n,k}(x, \vartheta)$ these conditions are identical for every k, but for each subclass the differential equations differ in the domain of λ, in accordance with (72).

Let us consider the condition at the left-hand end for $V_{n,k}(x, \vartheta)$ with fixed k different from 1 and $n-1$; let $\vartheta = \vartheta_0$ be the left-hand end. Then, putting $\vartheta = \vartheta_0$ in (73), we have $\lambda = \text{const.} = \rho_k^*$. In addition, from the previous relation $\Theta = \partial V_{n,k}(\rho_k^*, \vartheta)/\partial \vartheta$ we obtain

$$\left(\frac{\partial V_{n,k}(\rho_k^*, \vartheta)}{\partial \vartheta}\right)_{\vartheta = \vartheta_0} = 0.$$

In addition, we must take account of the condition

(74) $$\left(\frac{\partial^m V_{n,k}(x, \vartheta)}{\partial x^m}\right)_{x = \rho_k^*} = 0 \quad \text{for} \quad m = 1, 2, 3,$$

which follows from the qualitative nature of the deformation of the polynomials $V_{n,k}(x, \vartheta) = W_{n,k}(x, \Theta)$ explained on p. 105.

These conditions determine a numerical value for ϑ_0.

We note that the problem of integrating the system (73) with the given conditions is overdetermined in the sense that there are too many supplementary conditions; but since we already know that the polynomial $V_{n,k}(x, \vartheta)$ exists, the system and the conditions are consistent. The development of specific analytic or numerical methods for approximate integration of (73) is not part of our problem at present.

We shall now verify (73) and the supplementary conditions for some of the polynomials of Examples 38-44.

EXAMPLE 45. We had (Example 38)

$$V_{3,2}(x, \vartheta) = \vartheta x^3 - (2\sqrt{2\vartheta} + \vartheta)x^2 + 2(1 + \sqrt{2\vartheta})x - 1.$$

The resolvent of this polynomial is

$$x^3 - \left(1 + \sqrt{\frac{2}{\vartheta}}\right) x^2 + \sqrt{\frac{2}{\vartheta}}\, x.$$

It is clear that (71) is satisfied.

In addition, $\overline{V}_{3,2}(x) = 1 - 2(1 - x)^3 = 2x^3 - 6x^2 + 6x - 1$, i.e $\vartheta_0 = 2$ and the supplementary conditions (74) are also clearly satisfied.

We had

$$V_{4,2}(x, \vartheta) = \vartheta x^4 - 2\vartheta x^3 + (\vartheta + 2\sqrt{2\vartheta})x^2 - 2\sqrt{2\vartheta}x + 1$$

and

$$\overline{V}_{4,2}(x) = 32x^4 - 64x^3 + 48x^2 - 16x + 1.$$

Consequently conditions (71) and (74) are satisfied, since the resolvent of the polynomial $V_{4,2}(x, \vartheta)$ is

$$\begin{aligned} R_{4,2}(x) &= x(x - 1)(x - \rho)(x - \overline{1 - \rho}) \\ &= x^4 - 2x^3 + \left(1 + \sqrt{\frac{2}{\vartheta}}\right) x^2 - \sqrt{\frac{2}{\vartheta}}\, x \end{aligned}$$

(Example 39).

CHAPTER VI

POLYNOMIALS OF PASSPORT $[n, n-1, 0]$

§1. Determining functional and qualitative investigation

As we remarked on p. 79, the determining segment-functional of this passport can be taken in the form

$$(75) \qquad 0_0, 0_1, \cdots, 0_{n-3}, 1_{n-2}, \Theta_1, \Theta_2,$$

where (Θ_1, Θ_2) determines, in a two-dimensional region M in its domain and in a one-to-one way except for sign, precisely the polynomials of the specified passport. We denote them by $\{A_n(x, \Theta_1, \Theta_2)\}$.

By the corollary of Theorem 35, the segment (75) is served outside M by the polynomials with more than $n-1$ nodes; and $q > n-2$. As we already know, these polynomials have one of the passports $[n, n, 0]$, $[n, n, 1]$, with their subclasses. It is clear that in special cases polynomials of passport $[n, n+1, 0]$, i.e. $\pm T_n(x)$, may occur; also those of passport $[n-1, n, 0]$, i.e. $T_{n-1}(x)$; and, as we shall see, also $+T_{n-2}(x)$.

REMARK. The polynomial $+T_{n-2}(x)$ serves the segment only at the point $\Theta_1 = \frac{1}{2}(n-1)$, $\Theta_2 = n(2n-1)/16$. In fact, the basis of the segment (75), i.e. the segment $0_0, 0_1, \cdots, 0_{n-3}, 1_{n-2}$, is served by $T_{n-2}(x)$; consequently there is a single best extension $\Theta = \Theta_1^*$, $\Theta_2 = \Theta_2^*$.

It is defined by the resolvent of

$$T_{n-2}(x) = 2^{2n-5}x^{n-2} - (n-2)2^{2n-6}x^{n-3} + \frac{(n-2)(2n-7)}{2} \cdot 2^{2n-8}x^{n-4} - \cdots,$$

i.e. by

$$R_{n-1}(x) = \frac{T'_{n-2}(x)x(x-1)}{(n-2)\cdot 2^{2n-5}} = x^{n-1} - \frac{n-1}{2}x^{n-2} + \frac{2n^2-7n+4}{16}x^{n-3} - + \cdots$$

(p. 42). Clearly $-T_{n-2}(x)$ cannot serve the segment (75).

It is easily found that the polynomials $A_n(x, \Theta_1, \Theta_2)$ fall into subclasses, since they are obtained by deformation from the polynomials $\pm T_n(x)$

with the loss of two adjacent nodes (simultaneously or consecutively). Depending on the indices of the dropped nodes in $(\tau_i)_0^n$ we obtain n subclasses, and consequently $n+1$ subpassports $[n, n-1, 0(k, k+1)]$ $(k = 0, 1, \cdots, n-1)$ and $[n, n-1, 0(0, n)]$. This distinction will have essential significance later for obtaining boundary conditions for systems of differential equations. We use the term "polynomials with interior deformation" for extremal polynomials which are primitive and have all the roots of their first derivatives in the interval $[0, 1]$.

If at least one of these conditions is violated but the first derivative has no roots on $[0, 1]$ except nodes, we call the family a family with exterior deformation.

Thus the polynomials of passport $[n, n, 0]$ are typical polynomials with exterior deformation, and the subclasses of polynomials $W_{n,1(k)}(x)$ are examples of polynomials with interior deformation.

It is clear that we can find polynomials with mixed deformation that do not belong to either category; for example, $W_{n,1(k)}(\alpha x + \beta)$, the transformation of an arbitrary polynomial of passport $[n, n, 1(k)]$.

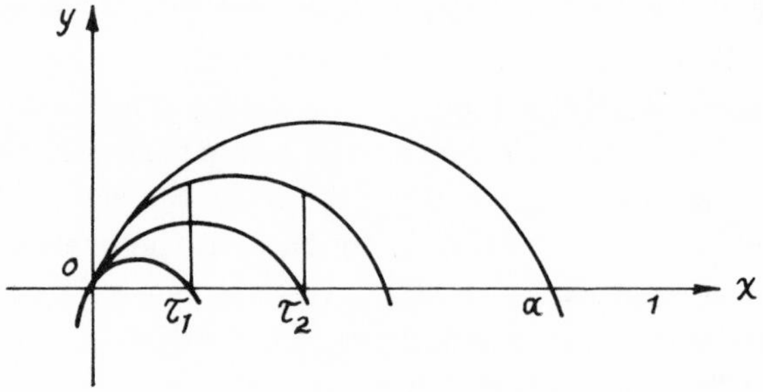

FIGURE 2.

We shall determine the region in the (Θ_1, Θ_2)-plane where the functional (75) is served by the polynomials $\pm T_n(x)$. Let Θ_1 be fixed; we find the corresponding values Θ_2' and Θ_2''. Decomposing (75) in terms of the nodes $(\tau_i)_0^n$, we have

$$\delta_k = (-1)^{n-k} \frac{\Theta_2 - \Theta_1 S_1^{(k)} + S_2^{(k)}}{\prod_{i \neq k} |\tau_k - \tau_i|}, \quad \text{where} \quad S_1^{(k)} = S_1 - \tau_k;$$

$$S_2^{(k)} = S_2 - \tau_k S_1 + \tau_k^2 \quad (k = 0, 1, 2, \cdots, n).$$

Here

$$S_1 = \sum_0^n \tau_i = \frac{n+1}{2}; \quad S_2 = \sum \tau_i \tau_j = \frac{2n^2 + n - 2}{16};$$

we have, according to our rule (p. 47), $\Theta_2 = (\Theta_1 - S_1S_2) + \tau_k(S_1 - \Theta_1 - \tau_k) = A + A_k$. We have to find the smallest and largest values of A_k $(k = 0, 1, \cdots, n)$. Put $\tau_k = x$ and $\frac{1}{2}(n+1) - \Theta_1 = \alpha$, and consider the family of parabolas $y = x(\alpha - x)$ on $[0,1]$ (Figure 2).

We form the following table:

$0 \leqq x \leqq 1$	α	$\leqq 0$	$0 < \alpha < 2$		$\alpha \geqq 2$
	Θ_1	$\geqq \dfrac{n+1}{2}$	$\dfrac{n-3}{2} < \Theta_1 < \dfrac{n+1}{2}$		$\leqq \dfrac{n-3}{2}$
	$y_{\max}$	$y(0) = 0$			$y(1) = \alpha - 1$
	$y_{\min}$	$y(1) = \alpha - 1$	$0 < \alpha \leqq 1$ $y(1) = \alpha - 1$	$1 \leqq \alpha < 2$ $y(0) = 0$	$y(0) = 0$

An analysis of $y_{\max}$ for $0 < \alpha < 2$ requires more detailed study since the set $x = \tau_k$ $(k = 1, 2, \cdots, n-1)$ is discrete.

We have

$$\Theta_2' = \frac{n+1}{2}\Theta_1 - \frac{2n^2 + n - 2}{16} + \min_{(k)} A_k$$

and

$$\Theta_2'' = \frac{n+1}{2}\Theta_1 - \frac{2n^2 + n - 2}{16} + \max_{(k)} A_k.$$

We have the following values for Θ_2' on the separate intervals for Θ_1:

$$\Theta_2' = \frac{n-1}{2}\Theta_1 - \frac{2n^2 - 7n + 6}{16} \quad \text{for} \quad \frac{n-1}{2} \leqq \Theta_1 < +\infty,$$

and the node $\tau_n = 1$ drops out;

$$\Theta_2' = \frac{n-1}{2}\Theta_1 - \frac{2n^2 + n - 2}{16} \quad \text{for} \quad -\infty < \Theta_1 \leqq \frac{n-1}{2},$$

and the node $\tau_0 = 0$ drops out.

These two rays starting from the point $((n-1)/2, (2n^2 - n - 2)/16)$ are shown in Figure 4 (p. 122), where we give the "topography" of the service of the segment (75) in the (Θ_1, Θ_2)-plane. At a general point of one of the rays, (75) is served by $-T_n(x)$ with two unweighted boundary nodes. Below these two rays the segment is served by $-T_n(x)$ with weights at all nodes.

We have, by the table and the formula for θ_2'':

$$\theta_2'' = \frac{n-1}{2}\theta_1 - \frac{2n^2-7n+6}{16} \quad \text{for } -\infty < \theta_1 \leqq \frac{n-3}{2},$$

and the node $\tau_n = 1$ drops out;

$$\theta_2'' = \frac{n+1}{2}\theta_1 - \frac{2n^2+n-2}{16} \quad \text{for } \frac{n+1}{2} \leqq \theta_1 < +\infty,$$

and the node $\tau_0 = 0$ drops out.

Consequently (Figure 4) the lower left-hand boundary ray becomes the upper right-hand boundary ray, and conversely.

We still have to determine the nature of the upper boundary in the interval $\frac{1}{2}(n-3) < \theta_1 < \frac{1}{2}(n+1)$.

We return to the study of the maxima of the parabolas $y = x(\alpha - x)$ for $0 < \alpha < 2$ under the condition that the admissible values for x are $(\tau_i)_0^n$:

$$A_k = \tau_k\left(\frac{n+1}{2} - \theta_1\right) - \tau_k^2 \quad (k = 0, 1, \cdots, n);$$

$$\theta_2 = \left(\frac{n+1}{2} - \tau_k\right)\theta_1 + \tau_k\left(\frac{n+1}{2} - \tau_k\right) - \frac{2n^2+n-2}{16},$$

a line with slope $\frac{1}{2}(n+1) - \tau_k$;

$x = \frac{1}{2}\alpha$ is the maximum point (Figure 3).

We form the following table:

α	$0 < \alpha \leqq \tau_1$	$\tau_1 \leqq \alpha \leqq \tau_1 + \tau_2$	$\tau_1 + \tau_2 \leqq \alpha \leqq \tau_2 + \tau_3$	etc.	$\tau_{n-1} + 1 \leqq \alpha < 2$
θ_1	$\frac{n+1}{2} - \tau_1 \leqq \theta_1 \leqq \frac{n+1}{2}$	$\frac{n+1}{2} - \tau_1 - \tau_2 \leqq \theta_1 \leqq \frac{n+1}{2} - \tau_1$	$\frac{n+1}{2} - \tau_2 - \tau_3 \leqq \theta_1 \leqq \frac{n+1}{2} - \tau_1 - \tau_2$		$\frac{n-3}{2} \leqq \theta_1 \leqq \frac{n-1}{2} - \tau_{n-1}$
max	$y(0) = 0$	$y(\tau_1)$	$y(\tau_2)$		$y(\tau_n) = y(1)$

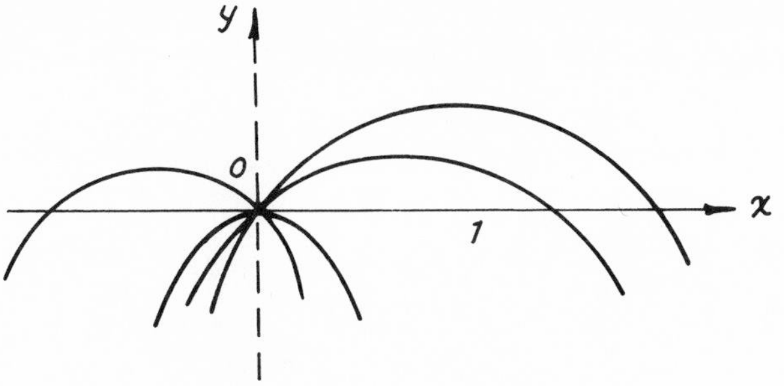

FIGURE 3.

When $\alpha/2 = (\tau_p + \tau_{p+1})/r$, the maximum of A_k is $y(\tau_p) = y(\tau_{p+1})$. Therefore when Θ_1 increases from $\frac{1}{2}(n-3)$ to $\frac{1}{2}(n+1)$ we have, remembering that $1 - \tau_{n-k} = \tau_k$, the following stages:

1) for $\frac{1}{2}(n-3) \leqq \Theta_1 \leqq \frac{1}{2}(n-3) + \tau_1$ the slope of the boundary line is $\frac{1}{2}(n+1) - 1 = \frac{1}{2}(n-1)$, and on the line the node $\tau_n = 1$ drops out;

2) for $\frac{1}{2}(n-3) + \tau_1 \leqq \Theta_1 \leqq \frac{1}{2}(n-3) + \tau_1 + \tau_2$ the slope of the boundary line is $\frac{1}{2}(n-1) + \tau_1$, and on the line the node τ_{n-1} drops out;

. . . .

$n+1$) for $\frac{1}{2}(n-3) + \tau_{n-1} + 1 \leqq \Theta_1 \leqq \frac{1}{2}(n+1)$ the slope of the boundary line is $\frac{1}{2}(n-1) + 1 = \frac{1}{2}(n+1)$, and on the line the node $\tau_0 = 0$ drops out.

At the vertices of the highest broken line, pairs of nodes are lost in the following order from left to right:

$$(\tau_n, \tau_{n-1}), (\tau_{n-1}, \tau_{n-2}), \cdots, (\tau_1, \tau_0).$$

Thus the regions in which the segment (75) is served by the polynomials $+T_n(x)$ and $-T_n(x)$ lie above the upper broken line and below the lower one (both regions including the boundary).

We now investigate the strip between the two broken lines.

Here the serving polynomials have $s < n+1$ nodes.

We shall find the lines of best continuation, i.e. for each Θ_1 we find the corresponding Θ_2^*.

By Example 34, for the truncated segment $0_0, 0_1, \cdots, 0_{n-3}, 1_{n-2}, \Theta_1$ we have the critical interval from $\Theta_1' = \frac{1}{2}(n-2)$ to $\Theta_1'' = \frac{1}{2}n$, outside which the extremal polynomial is $\pm T_n(x)$. The best continuation of

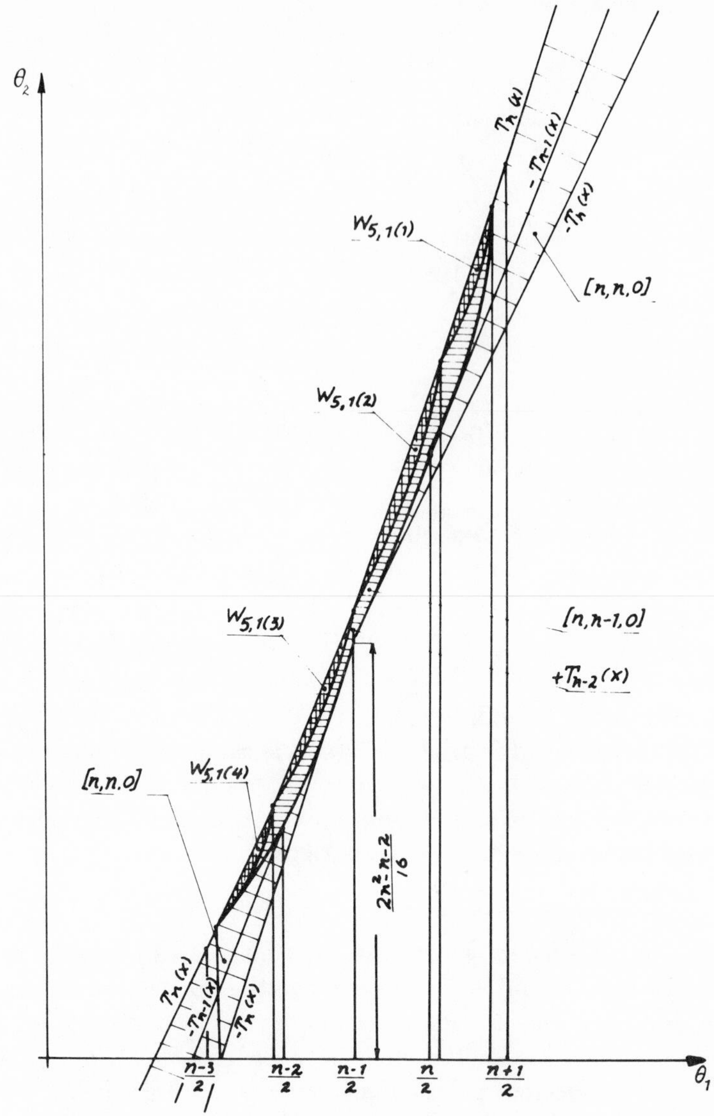

FIGURE 4. Topography of service of the functional by polynomials of passport $[n, n-1, 0]$ for $n = 5$

the segment, preserving these extremal polynomials, is found with the aid of the resolvent of $\pm T_{n-1}(x)$:

$$R_n(x) = x^n - S_1^{(n-1)}x^{n-1} + S_2^{(n-1)}x^{n-2} - + \cdots.$$

Then we have

$$\Theta_2^* - S_1^{(n-1)}\Theta_1 + S_2^{(n-1)} = 0; \tag{76}$$

since $S_1^{(n-1)} = n/2$ and $S_2^{(n-1)} = (2n^2 - 3n - 1)/16$ (found from S_1 and S_2 for T_n by replacing n by $n - 1$), we have directly that (76) is served by $-T_{n-1}(x)$ (Figure 4). We have

$$\Theta_2 = \frac{n}{2}\Theta_1 - \frac{2n^2 - 3n - 1}{16},$$

but in the two separate intervals $-\infty < \Theta_1 \leqq \frac{1}{2}(n-2)$ and $\frac{1}{2}n \leqq \Theta_1 < +\infty$ the serving polynomials are $-T_{n-1}(x)$ and $+T_{n-1}(x)$, respectively; at the boundary points $\frac{1}{2}(n-2)$ and $\frac{1}{2}n$ the node $\tau_0^{(n-1)} = 0$ or $\tau_{n-1}^{(n-1)} = 1$ loses its weight, according to what was proved in Chapter 4. Therefore in any vertical direction, i.e. for $\Theta_1 =$ const. and not in the interval $(\frac{1}{2}(n-3), \frac{1}{2}(n+1))$, as Θ_2 varies from Θ_2' to Θ_2'', service passes to polynomials of passport $[n,n,0]$. Thus the required region M in which (75) is served by the polynomials $A_n(x,\Theta_1,\Theta_2)$ certainly lies in a bounded region bounded on the left and right by the lines $\Theta_1 = \frac{1}{2}(n-3)$ and $\Theta_1 = \frac{1}{2}(n+1)$.

We now consider the intersection of this strip with the line $\Theta_1 = \frac{1}{2}(n-1)$.

We found that in this case $\Theta_2' = (2n^2 - n - 2)/16$ and that $-T_n(x)$ loses its weights at $\tau_0 = 0$ and $\tau_n = 1$ simultaneously. Then as Θ_2 increases still further the serving polynomial has to be a bilateral Čebyšev transformation $-T_n(\alpha x + \beta)$ with $n - 1$ nodes; this polynomial, under exterior deformation, becomes $+T_{n-2}(x)$ when $\Theta_2 = (2n^2 - n)/16$, and as Θ_2 increases still further it becomes a polynomial with interior deformation, which for $\Theta_2 = \Theta''$ becomes $+T_n(x)$.

§2. Equations of the boundary of the region M

The boundary of $M(\Theta_1,\Theta_2)$, the region where $[n, n-1, 0]$ polynomials serve, is as follows:

1) a lower curve on which service is by $[n,n,0]$ polynomials with one omitted node, which must be an extremum (for preservation of alternation); 2) upper curves on which service is by polynomials of passport $[n,n,1]$, of all subclasses. We consider these cases separately.

1) Let $Q_n(x, \vartheta)$ be the complete (one-parameter) family of passport $[n, n, 0]$, i.e. the Zolotarev polynomials and the semiprimitive Čebyšev transformations.

Let $Q_n(x)$ be one of these polynomials with nodes $(\sigma_i)_1^n$ forming a distribution with total alternation; assume for definiteness $\overset{+}{\sigma}_n$. Decomposing the truncated segment $0_0, \cdots, 0_{n-3}, 1, \Theta_1$ in terms of these nodes, we have

$$\delta_k = (-1)^{n-k} \frac{\Theta_1 - S_1^{(k)}}{\prod_{i \neq k} |\sigma_k - \sigma_i|} \quad (k = 1, 2, \cdots, n),$$

where $S_1^{(k)} = \sum_{i \neq k} \sigma_i$. Therefore the condition that the whole segment is served requires that $\Theta_1 \geqq S_1^{(k)}$ and that Θ_2 is determined. In fact, if $R_n(x) = x^n - S_1 x^{n-1} + S_2 x^{n-2} - \cdots$ is the resolvent of $Q_n(x)$ we must have $\Theta_2 - S_1 \cdot \Theta_1 + S_2 = 0$.

Thus for $\Theta_1 \geqq S_1$ ($S_1^{(k)} = \max$ when $k = 0$) (75) serves the given segment on the ray $\Theta_2 = S_1 \Theta_1 - S_2$. When $\Theta = S_1$ the weight at the node σ_1 drops out; this node is on the boundary of M. (An analogous situation arises in the $\bar{\sigma}_n$ case, with $\Theta_1 \leqq S_1 - \sigma_n$.) If we now replace the fixed $Q_n(x)$ by $Q_n(x, \vartheta) = \vartheta x^n + \cdots$ with a variable ϑ we obtain the lower boundary of the region of service, the curve with the equation $\Theta_2(\vartheta) - S_1(\vartheta) \cdot \Theta_1 + S_2(\vartheta) = 0$.

2) In contrast to the lower boundary of M, on which one or two extreme nodes drop out, and consequently inside M service is given by polynomials with exterior deformation, on the upper Čebyšev broken line either one or two interior nodes drop out.

Let us consider the possibility that the segment (75) is served by one-parameter polynomials of passport $[n, n, 1(k)]$ belonging to the kth subclass. Let $W_n(x)$ be a particular such polynomial with nodes $(\sigma_i)_1^n$ and suppose that its distribution contains $(\overset{+}{\sigma}_k, \overset{+}{\sigma}_{k+1})$.

We decompose the segment

$$0_0, 0_1, \cdots, 0_{n-3}, 1_{n-2}, \Theta_1 \tag{77}$$

in terms of this set of nodes; we have

$$\delta_m = (-1)^{n-m} \frac{\Theta_1 - S_1^{(m)}}{\prod_{i \neq m} |\sigma_m - \sigma_i|} \quad (m = 1, 2, \cdots, n).$$

Since $S_1 - 1 = S_1^{(n)} < S_1^{(n-1)} < \cdots < S_1^{(k+1)} < S_1^{(k)} < \cdots < S_1^{(2)} < S_1^{(1)} = S_1$ ($\sigma_1 = 0$, $\sigma_n = 1$), a necessary and sufficient condition for a permanence of sign $\delta_k > 0$, $\delta_{k+1} > 0$ is that

(78) $$S_1^{(k+1)} < \Theta_1 < S_1^{(k)},$$

and the condition that the whole segment (75) is served requires further that

(79) $$\Theta_2 - S_1 \cdot \Theta_1 + S_2 = 0$$

(as is immediately obtainable from the resolvent of W_n). Thus $W(x)$ serves (75) on the segment of the line (79) in the interval (78). The regions in which all the $\{W_{n,1(k)}(x)\}$ serve are conventionally represented as crescents in Figure 4. In the limit when $S_1^{(k+1)}(\vartheta) = S_1^{(k)}(\vartheta)$ the line segment reduces to a point.

Let us find the equation of the lower boundary of such a crescent.

Since the basis of the segment (75), $0_0, 0_1, \cdots, 0_{n-3}, 1_{n-2}$, has constant decomposition with respect to any $n-1$ nodes, we can decompose it with respect to $(\sigma_i)_{i \neq k}$ [or $(\sigma_i)_{i \neq k+1}$]. Put

$$\delta_m = (-1)^{n-m} \cdot \frac{1}{\prod |\sigma_m - \sigma_i|} \qquad (i \neq m;\ m, i \neq k)$$

$(m = 1, 2, \cdots, k-1, k+1, \cdots, n)$; $\sum_{i \neq k} \delta_i \sigma_i^m = 0$ for $m = 0, 1, \cdots, n-3$;

$$\sum_{i \neq k} \delta_i \sigma_i^{n-2} = 1; \quad \Theta_1 = \sum_{i \neq k} \delta_i \sigma_i^{n-1}; \quad \Theta_2 = \sum_{i \neq k} \delta_i\, \sigma_i^n.$$

Therefore we obtain unique values for Θ_1, Θ_2, which means that $W_n(x)$ with an unweighted node serves the segment (75) at two and only two points; one was (Θ_1, Θ_2) which we just found; the other is obtained for $i \neq k-1$.

The points we have found necessarily lie on the boundary of M, since one of the nodes of $W_n(x)$ ($\delta_k = 0$ or $\delta_{k+1} = 0$) was not used; consequently the passport of the segment is $[n, n-1, 0]$.

The equation of the curve where the $W_{n,1(k)}(x)$ serve—this curve is part of the upper boundary of M—can be found by considering $W_n(x, \vartheta) = \vartheta x^n + y_{n-1}(\vartheta) x^{n-1} + y_{n-2}(\vartheta) x^{n-2} + \cdots$ instead of $W_n(x)$. Parametric equations for the curve are

$$\Theta_1(\vartheta) = \sum_{i \neq k} \delta_i [\sigma_i(\vartheta)]^{n-1}; \quad \Theta_2(\vartheta) = \sum_{i \neq k} \delta_i [\sigma_i(\vartheta)]^n.$$

It is convenient to introduce the resolvent of $W_n(x, \vartheta)$:

$$R_n(x) = x^n - S_1(\vartheta) x^{n-1} + S_2(\vartheta) x^{n-1} - \cdots,$$

and then we have

$$\Theta_2(\vartheta) - S_1(\vartheta)\Theta_1 + S_2(\vartheta)\cdot 1 = 0.$$

We complete the qualitative study of the polynomials $A_n(x, \Theta_1, \Theta_2)$ by determining the curve passing through M on which the degree of the polynomial is decreased. This curve, as we have already shown, passes through the point

$$\left(\frac{n-2}{2}, \frac{2n^2-n}{16}\right).$$

In addition, polynomials of passport $[n-1, n-1, 0]$ belong to the set of $A_n(x, \Theta_1, \Theta_2)$; let us denote them by $Q_{n-1}(x, \Theta_1)$. We recall that their determining functional can be taken in the form $0_0, 0_1, \cdots, 0_{n-3}, 1_{n-2}, \Theta_1$; the corresponding Θ_2 is determined from the resolvent $R_{n-1}(x) = x^{n-1} - S_1(\Theta_1)x^{n-2} + S_2(\Theta)x^{n-3} - \cdots$. That is, we have $\Theta_2 = S_1(\Theta_1) - S_2(\Theta_1)$ with Θ_1 inside the critical interval of the segment (77), i.e. $\frac{1}{2}(n-2) < \Theta_1 < \frac{1}{2}n$. This is the curve where the $Q_{n-1}(x, \Theta_1)$ serve; clearly the value found here is $\Theta_2 = \Theta_2^*$.

§3. Analytic construction of $[n, n-1, 0]$ polynomials

THEOREM 52. *The $n-2$ segment-functionals*

$$(\mu_i)_0^n = 0_0, 0_1, \cdots, 0_{n-3}, 1_{n-2}, \Theta_1, \Theta_2;$$

$$(80)\qquad (\mu_i')_0^n = 0_0, 0_1, \cdots, -1, 0_{n-2}, \Theta_1^{(1)}, \Theta_2^{(1)};$$

$$\cdots\cdots\cdots\cdots\cdots\cdots\cdots\cdots$$

$$(\mu_i^{n-3}) = 0_0, (-1)^{n-3}, \cdots, 0_{n-3}, 0_{n-2}, \Theta_1^{n-3}, \Theta_2^{n-3}$$

form a system of equivalent segments, i.e. each of them is determining for the polynomials of passport $[n, n-1, 0]$.

In fact, the number of nodes of each segment is no less than $n-1$ (Theorem 26). Furthermore, a basis for any one of them has the form $0_0, \cdots, 0_{k-1}, (-1)^{n-k}, 0_{k+1}, \cdots, 0_{n-2}$, and has the invariant decomposition property with respect to any $(\rho_i)_1^{n-1}$ from the interval $[0,1]$ with $\rho_i \neq \rho_j$, where the weights have alternating sign. This is immediate if we take into account that the segment equal to this basis is served by one of the polynomials $\pm T_{n-2}(x)$, i.e. if it is decomposed in terms of the nodes $(\tau_i^{(n-2)})_{i=0}^{n-2}$ of these polynomials, it has weights of alternating sign also for any other location of the nodes on $[0,1]$.

Thus each segment $(\mu_i^{(k)})$ is determining for $[n, n-1, 0]$ polynomials in a certain two-dimensional region M_k. We denote this as follows:

$$\{A_n(x,\Theta_1,\Theta_2)\} \equiv \{A_n(x,\Theta_1^{(1)},\Theta_2^{(1)}\} \equiv \cdots \equiv A_n(x,\Theta_1^{(n-3)},\Theta_2^{(n-3)}).$$

Taking as fundamental the parameters Θ_1 and Θ_2 already used for the first segment, we introduce the corresponding parameters $\Theta_1^{(l)}, \Theta_2^{(l)}$ for the other segments. Put $A_n(x,\Theta_1,\Theta_2) = \sum_0^n a_i(\Theta_1,\Theta_2)x^i$ and note that the remark made on p. 92 in the similar case of constructing the $[n,n,0]$ polynomials is still valid here. Here also if we suppose that $A_n(x,\Theta_1,\Theta_2)$ is differentiable with respect to both parameters, we obtain the following condition on $(\mu_i^{(l)})$ for fixed $\Theta_1^{(l)}$ and $\Theta_2^{(l)}$:

$$a_n\Theta_2^{(l)} + a_{n-1}\Theta_1^{(l)} + (-1)^l a_{n-l-2} = \max(\Theta_1,\Theta_2).$$

Consequently we have formally

$$(81)\qquad \begin{aligned} &\frac{\partial a_n}{\partial\Theta_1}\Theta_2^{(l)} + \frac{\partial a_{n-1}}{\partial\Theta_1}\Theta_1^{(l)} + (-1)^l\frac{\partial a_{n-l-2}}{\partial\Theta_1} = 0;\\ &\frac{\partial a_n}{\partial\Theta_2}\Theta_2^{(l)} + \frac{\partial a_{n-1}}{\partial\Theta_2}\Theta_1^{(l)} + (-1)^l\frac{\partial a_{n-l-2}}{\partial\Theta_2} = 0. \end{aligned}$$

In order to form differential equations satisfied by the coefficients $a_i(\Theta_1,\Theta_2)$, it is convenient to replace Θ_1 and Θ_2 by other parameters.

THEOREM 53. *In the family of polynomials* $A_n(x,\Theta_1,\Theta_2)$ *of passport* $[n,n-1,0]$ *the parameters* $(\Theta_1,\Theta_2)\in M$ *can be replaced by equivalent parameters* $(\vartheta_1,\vartheta_2)$, *where* ϑ_1 *and* ϑ_2 *are the leading coefficients of the polynomial*:

$$\vartheta_2x^n + \vartheta_1x^{n-1} + y_{n-2}(\vartheta_1,\vartheta_2)x^{n-2} + \cdots + y_0(\vartheta_1,\vartheta_2),$$

and the $y_i(\vartheta_1,\vartheta_2)$ *are differentiable functions of* ϑ_1 *and* ϑ_2.

Since the segment (75) with a given pair of numbers (Θ_1,Θ_2) completely determines $Q_n(x)$, it follows that $\vartheta_1 = \phi_1(\Theta_1,\Theta_2)$ and $\vartheta_2 = \phi_2(\Theta_1,\Theta_2)$ are well-defined functions.

We have to prove that conversely $\Theta_1 = \Psi_1(\vartheta_1,\vartheta_2)$ and $\Theta_2 = \Psi_2(\vartheta_1\ \vartheta_2)$, i.e. each polynomial of passport $[n,n-1,0]$ is completely determined by its two leading coefficients.

Suppose that there are two polynomials of the form

$$Q_n^{(1)}(x) = q_0x^n + q_1x^{n-1} + q_2x^{n-2} + \cdots$$

and

$$Q_n^{(2)}(x) = q_0x^n + q_1x^{n-1} + \bar{q}_2x^{n-2} + \cdots,$$

where $\bar{q}_2 \neq q_2$. Suppose that the first serves the functional (75) for $(\Theta_1^{(1)}, \Theta_2^{(1)})$, and the second, for $(\Theta_1^{(2)}, \Theta_2^{(2)})$. Then, letting α denote either of $q_2, \bar{q}_2$, we have

$$[q_0\Theta_2^{(1)} + q_1\Theta_1^{(1)}] + \alpha = \max(\alpha); \quad [q_0\Theta_2^{(2)} + q_1\Theta_1^{(2)}] + \alpha = \max(\alpha);$$

since the terms in brackets are constants, α must be the same in both cases, namely the larger of the two assumed numbers q_2, $\bar{q}_2$.

Thus α is unique.

In the same way we can prove that q_0, q_1 determine q_3 etc. up to q_{n-1} uniquely; we have to use other functionals from (80).

Therefore the data q_0 and q_1 determine the remaining coefficients, except perhaps the constant term q_n. But if the constant terms in $Q_n^{(1)}(x)$ and $Q_n^{(2)}(x)$ are different but the other coefficients are equal, we have $Q_n^{(2)}(x) \equiv Q_n^{(1)}(x) + \text{const.}$, which is impossible for two reduced polynomials. Then $\Theta_1 = \Psi_1(\vartheta_1, \vartheta_2)$ and $\Theta_2 = \Psi_2(\vartheta_1, \vartheta_2)$ are the unique inverse functions.

Thus we have

$$\left.\begin{aligned} \Theta_1^{(l)} &= \Psi_1(\vartheta_1, \vartheta_2) \\ \Theta_2^{(l)} &= \Psi_2(\vartheta_1, \vartheta_2) \end{aligned}\right\} * \qquad (l = 0, 1, 2, \cdots, n-3)$$

and $A_n(x, \Theta_1^{(l)}, \Theta_2^{(l)}) \equiv \bar{A}_n(x, \vartheta_1, \vartheta_2)$.

We now show that $y_l(\vartheta_1, \vartheta_2)$ are differentiable.

The relations $*$ can be found by a different method. We have

$$\vartheta_2\Theta_2^{(l)} + \vartheta_1\Theta_1^{(l)} + (-1)^l y_l(\vartheta_1, \vartheta_2) = \max(\vartheta_1, \vartheta_2),$$

or, assuming the differentiability of $y_l(\vartheta_1, \vartheta_2)$,

$$\Theta_1^{(l)} + (-1)^l \frac{\partial y_l}{\partial \vartheta_1} = 0; \quad \Theta_2^{(l)} + (-1)^l \frac{\partial y_l}{\partial \vartheta_2} = 0.$$

Hence Ψ_1, Ψ_2 can be expressed uniquely in terms of $\partial y_l/\partial\vartheta_1$, $\partial y_l/\partial\vartheta_2$; therefore these derivatives exist and are continuous (by the theorem on continuous deformation).

Thus $\bar{A}_n(x, \vartheta_1, \vartheta_2)$ is a polynomial that is differentiable with respect to ϑ_1 and ϑ_2.

Let its resolvent be

$$R_{n-1}(x, \vartheta_1, \vartheta_2) = x^{n-1} + \alpha_{n-2}(\vartheta_1, \vartheta_2)x^{n-2} + \cdots + \alpha_0(\vartheta_1, \vartheta_2).$$

THEOREM 54. *The following differential relationships connect* $\bar{A}(x, \vartheta_1, \vartheta_2)$ *and* $R_{n-1}(x, \vartheta_1, \vartheta_2)$:

$$\text{(82)} \qquad \frac{\partial \bar{A}_n(x, \vartheta_1, \vartheta_2)}{\partial \vartheta_1} = R_{n-1}(x, \vartheta_1, \vartheta_2);$$

$$(83) \qquad \left(x - \frac{\partial y_{n-2}}{\partial \vartheta_1}\right) \cdot R_{n-1}(x, \Theta_1, \Theta_2) = \frac{\partial \overline{A}_n(x, \vartheta_1, \vartheta_2)}{\partial \vartheta_2}.$$

In fact, by the properties of the resolvent we have the following two sets of equations for each of the equivalent segments (80):

$$(\mathrm{I}) \qquad \begin{cases} \Theta_1 + \alpha_{n-2} = 0 \\ \Theta_1^{(1)} - \alpha_{n-3} = 0 \\ \cdots\cdots\cdots\cdots \\ \Theta_1^{(n-3)} + (-1)^{n-3}\alpha_1 = 0 \end{cases}$$

$$(\mathrm{II}) \qquad \begin{cases} \Theta_2 + \alpha_{n-2}\Theta_1 + \alpha_{n-3} = 0 \\ \Theta_2^{(1)} + \alpha_{n-2}\Theta_1^{(1)} - \alpha_{n-4} = 0 \\ \cdots\cdots\cdots\cdots\cdots\cdots \\ \Theta_2^{(n-3)} + \alpha_{n-2}\Theta_1^{(n-3)} + (-1)^{n-3}\alpha_0 = 0 \end{cases}$$

From equations (I) we immediately have

$$\alpha_{n-2} = -\Theta_1 = \frac{\partial y_{n-2}}{\partial \vartheta_1}; \; \alpha_{n-3} = \frac{\partial y_{n-3}}{\partial \vartheta_1}; \; \cdots; \; \alpha_0 = \frac{\partial y_0}{\partial \vartheta_1},$$

which yields (82). We get (83) from (II) by replacing the $\Theta_2^{(i)}$ and $\Theta_1^{(i)}$ by $\pm\, \partial y_i / \partial \vartheta_2$ and $\pm\, \partial y_i / \partial \vartheta_1$, multiplying successively by $x^{n-2}, x^{n-3}, \cdots, x$, and adding.

COROLLARY. *The polynomials $\overline{A}_n(x, \vartheta_1, \vartheta_2)$ satisfy the partial differential equation*

$$(84) \qquad \frac{\partial \overline{A}_n}{\partial \vartheta_2} = \left(x - \frac{\partial y_{n-2}}{\partial \vartheta}\right) \cdot \frac{\partial \overline{A}_n}{\partial \vartheta_1}.$$

Consequently the coefficients of the polynomial $y_i(\vartheta_1, \vartheta_2)$ satisfy a system of $n-1$ first-order partial differential equations. In addition to (82), (83), the polynomials $\overline{A}_n(x, \vartheta_1, \vartheta_2)$, if they are primitive, also satisfy the evident equation

$$(85) \qquad R_{n-1}(x, \vartheta_1, \vartheta_2) = \frac{\partial \overline{A}_n}{\partial x} \cdot \frac{(x-1)x}{n\vartheta_2(x-\gamma)(x-\delta)},$$

where γ and δ denote the two roots of $\partial \overline{A}_n / \partial x$ that do not appear among the nodes $(\sigma_i)_1^{(n-1)}$; these roots lie, depending on the subclass of the determining polynomial, either outside $[0, 1]$ or in an interval $[\tau_{k-1}, \tau_{k+2}]$ and can be expressed in terms of (y_i') in a way similar to that followed in Chapter IV.

After excluding $R_{n-1}(x)$, equations (85) and (82), (83) yield an overdetermined system for the coefficients, but it is always consistent. The boundary conditions can be obtained from the equations of the curves bounding the region M (Figure 4). These equations require a preliminary analytic or numerical integration for polynomials of higher passports, i.e. for $[n, n, 0]$ and $[n, n, 1(k)]$.

In concluding this chapter and the first part of the monograph we make some general remarks.

Under the assumption of the existence of a polynomial of given passport $[n, s, p]$ and the possibility of constructing a determining basis segment for every set of polynomials of this passport, we obtained the family $\{Q_n(x)\}$ depending on $n+1-s=l$ parameters (Theorems 32-34), where the parameters are independent, i.e. the whole family of polynomials can be written in the form $Q_n(x, \Theta_1, \Theta_2, \cdots, \Theta_l)$, where Θ_i are variable elements of the determining functional.

In addition, if we assume

1) that this system of parameters $(\Theta_i)_1^l$ can be replaced by a system of the l leading (or other) coefficients $(\vartheta_i)_1^l$, leaving the family still complete, i.e. if

$$\begin{aligned} \{Q_n(x, \Theta_1, \cdots, \Theta_l)\} \equiv \{ & \vartheta_l x^n + \vartheta_{l-1} x^{n-1} + \cdots + \vartheta_1 x^{n-l+1} \\ & + y_{n-l}(\vartheta_1, \cdots, \vartheta_l) x^{n-l} + \cdots + y_0(\vartheta_1, \cdots, \vartheta_l)\} \end{aligned} \tag{86}$$

and

2) that $y_i(\vartheta_1, \cdots, \vartheta_l)$ are differentiable with respect to each ϑ,

then the same system of differential equations can be obtained more simply than in Theorem 54, and moreover in the most general case, from the double structural identity (16):

$$Q_n(x, \vartheta_1, \vartheta_2, \cdots, \vartheta_l) = 1 - \prod (x - \sigma_i)^2 \phi(x) \equiv -1 + \prod (x - \sigma_i)^2 \psi(x), \tag{87}$$

by[1)] differentiating it formally with respect to the ϑ_i.

In (87) the possible factors x and $1-x$ are included respectively in $\phi(x)$ and $\psi(x)$; $(\overset{+}{\sigma}_i)$ and $(\bar{\sigma}_i)$ are functions of the coefficients $\vartheta_1, \vartheta_2, \cdots, \vartheta_l$. We have

$$\partial Q_n / \partial \vartheta_k = \Phi_1(x) \prod (x - \overset{+}{\sigma}_i) = \Phi_2(x) \prod (x - \bar{\sigma}_i);$$

hence because all the points $(\overset{+}{\sigma}_i)$ and $(\bar{\sigma}_i)$ are different, we have $\partial Q_n / \partial \vartheta_k = R_s(x) \Psi(x)$, i.e. the derivative of the extremal polynomial with respect to any ϑ_i is a multiple of the resolvent $R_s(x) = \prod (x - \overset{+}{\sigma}_i) \prod (x - \sigma_i)$.

1) This remark was made by E. L. Rabkin.

In particular, in view of the equation $n+1-l=s$ we have from (87)

$$\frac{\partial Q_n}{\partial \vartheta_1} = x^s + \frac{\partial y_{s-1}}{\partial \vartheta_1} x^{s-1} + \cdots$$

and consequently

$$\frac{\partial Q_n}{\partial \vartheta_1} = R_s(x);$$

$$\frac{\partial Q_n}{\partial \vartheta_2} = x^{s+1} + \frac{\partial y_{s-1}}{\partial \vartheta_2} x^{s-1} + \cdots, \text{ i.e. } \frac{\partial Q_n}{\partial \vartheta_2} = (x+a_2) R_s(x);$$

$$\frac{\partial Q_n}{\partial \vartheta_3} = x^{s+2} + \frac{\partial y_{s-1}}{\partial \vartheta_3} x^{s-1} + \cdots, \text{ i.e. } \frac{\partial Q_n}{\partial \vartheta_3} = (x^2 + a_3 x + b_3) R_s(x)$$

etc.

In these formulas the coefficients $a_2, a_3, b_3, \cdots$ are determined algebraically by setting certain coefficients in the derivatives $\partial Q / \partial \vartheta_i$ equal to zero; in fact, if we put $R_s(x) = x^s + \alpha_{s-1} x^{s-1} + \alpha_{k-2} x^{s-2} + \cdots$, we have the following conditions for determining $a_2, a_3, b_3, \cdots$:

$$a_2 + \alpha_{s-1} = 0; \quad a_3 + \alpha_{s-1} = 0; \quad b_3 + a_3 \alpha_{s-1} + \alpha_{s-2} = 0$$

etc.

Problems of integrating the overdetermined systems of differential equations for extremal polynomials are in general quite difficult if an approximate analytic solution is to be obtained.

This is not the case if we contemplate numerical integration and take advantage of the use of electronic machines. Thus in 1958 the Polish mathematician S. Paszkowski solved the problem for the simplest case considered in Chapter IV, i.e. for Zolotarev polynomials. If we replace the coefficients y_i in the system (60) by the nodes σ_k and put $t = 1/\lambda$, we obtain the canonical form

$$\frac{d\sigma_k}{dt} = \frac{\sigma_k(1-\sigma_k)}{(\sigma_k t - 1)\left[2 - t + (1-t) \sum_{i=2}^{n-1} (1-\sigma_i t)^{-t}\right]} \qquad (k = 2, 3, \cdots, n-1)$$

with the initial conditions that $\sigma_k = \tau_{k-1}/\cos^2(\pi/2n)$ at $t = 1$.

These equations are quite suitable for programming, after which the transition to the coefficients themselves is not difficult. Hence it is not necessary to use elliptic functions to determine the Zolotarev polynomials. Similarly we do not need automorphic functions for the polynomials $A_n(x_1, \vartheta_1, \vartheta_2)$ (Part 2, Chapter I).

Part Two

PROBLEMS OF ČEBYŠEV APPROXIMATION

Problems of best uniform approximation originate historically in the ideas and work of P. L. Čebyšev [**10**]. The first problem was this: Find a polynomial of degree not exceeding n that deviates least from zero on $[-1,1]$, with a given leading coefficient A. Čebyšev himself constructed as solution the famous polynomial $(A/2^{n-1})\cos n \arccos x$.

E. I. Zolotarev [**11**] set the analogous problem of finding such a polynomial when the two leading coefficients are given, and completely solved his own problem by constructing the required polynomial. However, the analytic form of the Zolotarev polynomials is rather complicated and involves elliptic functions.

V. A. Markov [**12**] proposed an extension of the problem: to find a polynomial deviating least from zero on a finite interval with a given linear inhomogeneous relation among the coefficients of the polynomial. He gave a constructive solution of his problem only in some special cases.

A paper [**13**] by A. P. Pšeborskiĭ is entirely in the spirit of Čebyšev, although it does not contain results of a constructive character. He studied the problem of the polynomial of least deviation under two linear relations among the coefficients.

It is clear that similar problems can be transformed in many ways. This has been done quite recently by various mathematicians, such as N. I. Ahiezer, D. G. Grebenjuk, G. M. Mirak'jan, and others.

It should be remarked that the study and actual construction (exact or approximate) of the required polynomial has been carried out from the beginning in each case, and for its own sake, i.e. the problems have been considered separately both as to formulation and as to method of solution.

The investigation of segment-functionals and their extremal polynomials shows the close connection of these polynomials with the solution of Čebyšev problems. The latter, as will be shown in Chapter I, can be transformed into the single problem of finding the extremal polynomial of a given segment-functional.

However, the value of the method of functionals is by no means exhausted by the simplest problems mentioned here; this will become clear in later chapters.

CHAPTER I

POLYNOMIALS OF LEAST DEVIATION FOR SOME CLASSICAL PROBLEMS

§1. V. A. Markov's problem

In the introduction to his memoir "On functions deviating least from zero on a given interval" (1892), V. A. Markov posed the following general problem.

In the domain of polynomials $\{P_n(x)\}$ of degree at most n with real coefficients, subject to the condition $\sum_{i=0}^{n} \alpha_i p_i = A$ $(a \neq 0)$ where $(p_i)_0^n$ are the coefficients of the polynomial, and $(\alpha_i)_0^n$ and A are given real numbers, find the polynomial deviating least from zero on a given interval $[a,b]$.

Letting $L_p = \max_{[a,b]} |P_n(x)|$, we seek a polynomial $Y_n(x)$ such that L_Y is as small as possible. We shall call this $Y_n(x)$ a polynomial of least deviation on $[a,b]$ for the system of numbers (α) and A.

First we show that two problems are the same: the problem of V. A. Markov just stated, and the problem of determining the principal polynomial $Q_n(x)$ of a given segment $(\mu_i)_0^n$.

THEOREM 55. *If we have a segment $(\mu_i)_0^n$ with principal polynomial $Q_n(x)$ and norm $N = Q_n(\bar{\mu})$, and a polynomial $Y_n(x)$ of least deviation L on $[0,1]$ for the system of numbers $(\mu_i)_0^n$ and $A > 0$, then*

$$Q_n(x) = Y_n(x)/L \quad \text{and} \quad N = A/L.$$

In fact, $Y_n(\bar{\mu})/L = A/L \leqq N$. But if $A/L < N$, then the deviation of the polynomial $(A/N)\,Q_n(x)$ on $[0,1]$ is $A/N < L$, which is impossible by hypothesis. Thus $A/L = N$.

Since the two problems are equivalent, it follows that to every theorem about one problem corresponds a theorem about the other. We note some evident corresponding results.

COROLLARY 1. *The polynomials $Q_n(x)$ and $Y_n(x)$ are (each for its own problem) simultaneously unique or not.*

It is obvious that if $(\overset{\pm}{\sigma}_i)_1^s$ is the distribution of $Q_n(x)$, then at each point $\overset{\pm}{\sigma}_i$ we have $Y_n(\overset{\pm}{\sigma}_i) = \pm L$. Conversely, if $(\rho_i)_1^s$ is the set of points where $Y_n(x) = \pm L$, then $Q_n(\rho_i) = \pm 1$. Thus the sets (σ_i) and (ρ_i) are the same.

COROLLARY 2. *If we replace A by $-A$ in V. A. Markov's problem, the required polynomial is $-Y_n(x)$, and $N = |A|/L$.*

Since whether $A > 0$ or $A < 0$ makes no difference in the problem of determining $Q_n(x)$, we obtain the following proposition by putting $A = N$: A principal polynomial $Q_n(x)$ for the segment $(\mu_i)_0^n$ is a polynomial of least deviation for the system $(\mu_i)_0^n$ and $A = N$, i.e. $Q_n(x) = Y_n(x)$. The degree of $Y_n(x)$ can be decreased if and only if $\mu_n = \mu_n^*$.

COROLLARY 3. *If the segment $(\mu_i)_0^n$ is of class* II, *then $Y_n(x)$ is unique, i.e. if $Y_n(x)$ is not unique the number s of its points of deviation satisfies $s \leqq \frac{1}{2}n + 1$.*

COROLLARY 4. *In V. A. Markov's problem, let the segment $(\mu_i)_0^n$ be absolutely monotonic; then*

1) *if $(\mu_i)_0^n$ has more than one best extension, then $Y_n(x)$ is unique and equal to a constant; in fact, since $N = \mu_0$ we have $Q_n(x) \equiv 1$ and $Y_n(x) = A/\mu_0$ $(A > 0)$;*

2) *if $(\mu_i)_0^n$ has a unique extension, then $Y_n(x)$ is not unique, $L = A/\mu_0$, and the common points of deviation, $s \leqq \frac{1}{2}n + 1$ in number, are located as in Theorem* 9;

3) *conversely, if $Y_n(x) =$ const., i.e. we have $Y_n(x) = +L$ at all points of deviation, then $(\mu_i)_0^n$ is absolutely monotonic.*

COROLLARY 5. *If at least one element μ_k $(k > 0)$ is outside or at an endpoint of its critical interval, and only if this occurs, V. A. Markov's problem has the solution $T_n(x) = \cos n \arccos(2x - 1)$; moreover,*

$$N = |T_n(\bar{\mu})|; \quad L = \frac{A}{|T_n(\bar{\mu})|} \ (A > 0); \quad Y_n(x) = LT_n(x),$$

where we are to take the $+$ or $-$ sign according to the indications of Theorem 18.

COROLLARY 6. *If the solution $Y_n(x)$ of Markov's problem is not unique (i.e., if $Q_n(x)$ is not unique), then among the solutions there is a polynomial that has not more than $\frac{1}{2}n + 1$ points of deviation (and there is an uncountable set of such polynomials); there are also solutions that have more than $\frac{1}{2}n + 1$ points of deviation. There are at least two of the latter (Theorem* 16).

COROLLARY 7. *Every polynomial $P_n(x)$ is a polynomial of least deviation for some Markov problem; that is, for every $P_n(x)$ we can find numbers $(\mu_i)_0^n$ and $A > 0$ such that this $P_n(x)$ is a polynomial of least deviation on $[0, 1]$ for the system of numbers $(\mu_i)_0^n$ and A.*

In fact, let $\max_{[0,1]}|P_n(x)| = L$ and let $(\sigma_i)_1^s$ be the points of deviation of $P_n(x)$ on $[0,1]$; then $Q_n(x) = P_n(x)/L$ has the distribution $(\overset{\pm}{\sigma_i})_1^s$. With this distribution, construct any segment $(\mu_i)_0^n$; then $Q_n(\overline{\mu}) = N$, where N is the norm of the segment. Clearly $P_n(x)$ is a polynomial of least deviation on $[0,1]$ for Markov's problem with the system of numbers $(\mu_i)_0^n$ and $A = NL$.

Since, when the segment $(\mu_i)_0^n$ is constructed, the moduli of the weights δ_i in the system $\sum_{i=1}^s \delta_i \sigma_i^k = \mu_k$ $(k = 0, 1, \cdots, n)$ are quite arbitrary, it follows that $P_n(x)$ is an extremal polynomial for an uncountable set of problems.

In the cited paper [12] Markov proves two necessary and sufficient criteria for the required polynomial of least deviation, and a theorem on necessary conditions for the uniqueness of the polynomial. We shall state some of his conclusions, using our terminology, and discuss them from the point of view of our theorems.

CRITERION 1. *A necessary and sufficient condition that a polynomial* $Q_n(x)$ *for which* $Q_n(\overline{\mu}) > 0$ *is an extremal polynomial for the segment* $(\mu_i)_0^n$ *is as follows: there is no polynomial* $\Psi_n(x)$ *for which simultaneously* $\Psi_n(\overline{\mu}) = 0$ *and* $\operatorname{sgn}\Psi(\sigma_i) = Q_n(\sigma_i)$ $(\neq 0)$, *where* $(\sigma_i)_1^s$ *are the nodes of* $Q_n(x)$.

NECESSITY. Let $Q_n(x)$ be an extremal polynomial for $(\mu_i)_0^n$. Then $\mu_k = \sum_{i=1}^s \delta_i \sigma_i^k$, and if $\Psi_n(x)$ exists we have $\Psi_n(\overline{\mu}) = \sum_1^s \delta_i \Psi_n(\sigma_i) = 0$ and $\operatorname{sgn}\Psi_n(\sigma_i) = \operatorname{sgn}\delta_i$ when $\delta_i \neq 0$; but then $\delta_i \Psi_n(\sigma_i) > 0$, and the first condition is impossible.

SUFFICIENCY. Suppose that there is no $\Psi_n(x)$ corresponding to $Q_n(x)$, with $\max|Q_n(x)| = 1$. We have to show that

$$|s_n(\overline{\mu})| \leqq Q_n(\overline{\mu}) \max_{[0,1]} |s_n(x)|$$

for every polynomial $s_n(x)$.

Let $Q_n(\overline{\mu}) = N$ $(N > 0$ by hypothesis$)$; suppose $s_n(\overline{\mu}) = N$, which is always possible if we multiply $s_n(x)$ by a suitable factor. Suppose $s_n(x) = Q_n(x) - \phi(x)$; then $\phi(\overline{\mu}) = 0$.

By hypothesis we do not have $\operatorname{sgn}\phi(\sigma_i) = Q_n(\sigma_i)$ for all σ_i, i.e. there is a point σ_k at which $|s_n(\sigma_k)| \geqq Q_n(\sigma_k) = 1$ (there is equality when $\phi(\sigma_k) = 0$). Putting $\max_{[0,1]}|s_n(x)| = L_s$, we have $L_s \geqq 1$. Consequently $s_n(\overline{\mu})/L_s = N/L_s \leqq N$, and N is the norm of the segment $(\mu_i)_0^n$.

CRITERION 2. *A necessary and sufficient condition that one of the polynomials* $\pm Q_n(x) \neq \text{const.}$ *is an extremal polynomial for the segment* $(\mu_i)_0^n$ *is as follows: If* $R(x) = \prod_1^s (x - \sigma_i)$ *is the resolvent of* $\pm Q_n(x)$ *and* $R^{(\sigma_k)}(x) = R(x)/(x - \sigma_k)$, *then* $\Phi(\overline{\mu}) = 0$ *for every polynomial* $\Phi(x)$ *of degree not exceeding n which is a multiple of the resolvent; the numbers*

$$\prod_k = (-1)^k R^{(\sigma_k)}(\bar{\mu})\, Q_n(\sigma_k),$$

must all have the same sign.

NECESSITY. Let $Q_n(x)$ be an extremal polynomial of the segment $(\mu_i)_0^n$. Let its nodes be $(\sigma_i)_1^s$. Then $\mu_k = \sum_{i=1}^s \delta_i \sigma_i^k$ $(k = 0, 1, \cdots, n)$, where some, but not all, δ_i can be zero, and the remaining δ_i satisfy $\operatorname{sgn} \delta_i = Q_n(\sigma_i)$.

We have

$$\Phi(\bar{\mu}) = \sum_{i=1}^s \delta_i \Phi(\sigma_i) = 0,$$

since $\Phi(\sigma_i) = 0$.

Let $(\rho_j)_1^{s_1}$ be the faithful nodes of the segment $(\mu_i)_0^n$; then $s_1 \leqq s$ and (ρ_i) is part of (σ_i). Thus $\delta_i = 0$ when σ_i differs from the ρ_j.

Further,

$$R^{(\sigma_k)}(\bar{\mu}) = \sum_{i=1}^s \delta_i R^{(\sigma_k)}(\sigma_i);$$

if σ_k is different from the ρ_j, we have $R^{(\sigma_k)}(\bar{\mu}) = 0$; if, however, $\sigma_k = \rho_l$ then $\sum_1^s$ reduces to a single term and $R^{(\sigma_k)}(\bar{\mu}) = \delta_k R^{(\sigma_k)}(\sigma_k)$.

Thus either $\prod_k = 0$ or

$$\prod_k = (-1)^k \delta_k Q_n(\sigma_k)\, R^{(\sigma_k)}(\sigma_k);$$

but $R^{(\sigma_k)}(\sigma_k) = R'(\sigma_k)$ and therefore

$$\operatorname{sgn} \prod_k = (-1)^k \operatorname{sgn} R'(\sigma_k) \quad \text{or} \quad \prod_k = 0.$$

But the signs of $R'(\sigma_k)$ $(k = 1, 2, \cdots, s)$ alternate, and consequently the assertion is proved.

SUFFICIENCY. Let there be given a segment $(\mu_k)_0^n$ and a polynomial $Q_n(x)$ with $\max_{[0,1]}|Q_n(x)| = 1$, and let the distribution of the polynomial be $(\overset{\pm}{\sigma}_i)_1^s$. It is known that $\Phi(\bar{\mu}) = 0$ if $\Phi(x)$ is any polynomial, of degree not exceeding n, that is a multiple of the resolvent $R(x) = \prod_1^s (x - \sigma_i)$. The numbers $(-1)^k R^{(\sigma_k)}(\bar{\mu}) Q_n(\sigma_k)$ are, for $k = 1, 2, \cdots, s$, of constant sign or zero (but not all zero).

We form the system of $n + 1$ equations in the s unknowns δ_i:

$$\sum_{i=1}^s \delta_i \sigma_i^k = \mu_k \qquad (k = 0, 1, \cdots, n). \tag{88}$$

Take $s < n + 1$ and find (δ_i) from the first s equations.

Put $\operatorname{sgn}\delta_i = (-1)^{s-i}\cdot\operatorname{sgn}R^{(\sigma_i)}(\bar{\mu})$ $(i = 1,2,\cdots,s)$; but by the second hypothesis, we also have that the numbers $(-)^{s-i}R^{(\sigma_i)}(\bar{\mu})\cdot Q_n(\sigma_i)$ are of constant sign or zero for $i = 1,2,\cdots,s$. Consequently for $i = 1,2,\cdots,s$ the numbers $\operatorname{sgn}\delta_i\cdot\operatorname{sgn}Q_n(\sigma_i)$ are of constant sign, or zero.

Choosing a suitable one of the polynomials $\pm Q_n(x)$, say $+Q_n(x)$, we take $\operatorname{sgn}\delta_i = \operatorname{sgn}Q_n(\sigma_i)$ or zero (for $i = 1,2,\cdots,s$).

In addition, by the first hypothesis,

$$R_s[\bar{\mu}] = 0;\ [xR_s(x)]_{\bar{\mu}} = 0;\ \cdots,\ [x^{n-s}\cdot R_s(x)]_{\bar{\mu}} = 0,$$

where $[x^iR(x)]_{\bar{\mu}} = F[x^iR(x)]$ is the functional defined by $(\mu_i)_0^n$, i.e. the whole segment $(\mu_i)_0^n$ is obtained from $(\mu_i)_0^{s-1}$ by extending it according to the resolvent $R(x)$ (p. 56).

Consequently the whole system (88) is consistent, the norm of the segment is $N \leqq \sum_{i=0}^n|\delta_i|$ (Theorem 13), and $Q_n(\bar{\mu}) = N$. If $s = n+1$, the first hypothesis of the consistency of the overdetermined system clearly drops out. This completes the proof of Criterion 2.

REMARK. After the investigation in Chapter 1 of this monograph we can formulate the following necessary and sufficient criterion for Markov's problem.

A necessary and sufficient condition that a given polynomial $L_n(x) \neq \text{const.}$ *deviates least from zero on* $[0,1]$ *among polynomials* $P_n(x) = \sum_0^n p_ix^i$ *whose coefficients satisfy* $\sum_0^n p_i\mu_i = A$ $(\neq 0)$ *is as follows: if* $\max_{[0,1]}|L_n(x)| = M > 0$ *and* $(\sigma_i)_1^s$ *are the points of* $[0,1]$ *at which* $L_n(\sigma_i) = \pm M$, *then the system of* $n+1$ *equations in* s *unknowns* δ_i,

$$\sum_{i=1}^{s}\delta_i\sigma_i^k = \mu_k \quad (k = 0,1,\cdots,n)$$

is consistent; $\operatorname{sgn}\delta_i = L_n(\sigma_i)/M$ *or* $\delta_i = 0$ $(i = 1,2,\cdots,s)$ (*but not all* δ_i *are zero*).

§2. The Zolotarev-Pšeborskiĭ problem

In his memoir "On a problem of smallest values" (1868), E. I. Zolotarev posed and solved the following problem: among polynomials $\{P_n(x)\}$, where $P_n(x) = \sum_0^n p_ix^i$, $p_n = 1$ and $p_{n-1} = -\sigma$ (σ is a given number), find $Y_n(x,\sigma)$ which deviates least from zero on the interval $(-1,+1)$.

Zolotarev established that these polynomials assume their maximum modulus exactly n times for $-1 \leqq x \leqq +1$, $0 < \sigma < \infty$, and that when $\sigma = 0$ we have $Y_n(x,0) = 2^{-n+1}\overline{T}_n(x)$, where $\overline{T}_n(x) = \cos n\arccos x$. Zolotarev found the deviation $L(\sigma)$, and also the polynomials $Y_n(x,\sigma)$ themselves, in terms of elliptic functions for $n\tan^2(\pi/2n) < \sigma < +\infty$

and in terms of trigonometric functions for $0 < \sigma < n \tan^2(\pi/2n)$.

It is useful to extend Zolotarev's problem by restating it as a problem formulated by A. P. Pšeborskiĭ [13].

PŠEBORSKIĬ'S PROBLEM. Among polynomials $\{P_n(x)\}$ whose coefficients satisfy two consistent conditions $\sum_0^n p_i \mu_i = A$ and $\sum_0^n p_i \nu_i = B$, find one that deviates least from zero on $[0, 1]$.

Let us denote the required polynomial by $L_n(x)$ and its deviation by

$$M_L = \max_{[0,1]} |L_n(x)|.$$

In the first place, it is clear that the Zolotarev-Pšeborskiĭ problem is equivalent in principle to finding the extremal polynomial for a certain segment $(\xi_i)_0^n$, i.e. it is equivalent to a Markov problem (Theorem 55). In fact, let $Q_n(x) = L_n(x)/M_L$; let $(\overset{\pm}{\sigma}_i)_1^s$ be the distribution of this polynomial; then if $(\xi_i)_0^n$ is any segment constructed with this whole distribution, $L_n(x)$ is a polynomial with least deviation among those whose coefficients satisfy

$$\sum_0^n p_i \xi_i = M_L \cdot Q_n(\bar{\xi}).$$

Therefore the two assumed relations among the coefficients of the required polynomial can always be reduced to one. Such a reduction is given in Theorem 56 under certain restrictive conditions.

THEOREM 56. *If $A \neq 0$ and $B \neq 0$ in Pšeborskiĭ's problem and if $Q_n(x, \Omega)$ is the family of extremal polynomials of the segment $\mu_0 + \Omega\nu_0$; $\mu_1 + \Omega\nu_1$; $\cdots$; $\mu_n + \Omega\nu_n$ with $-\infty < \Omega < +\infty$, and $L_n(x)$ with deviation M_L is the required polynomial, then a necessary and sufficient condition that $L_n(x)/M_L$ belongs to the family $Q_n(x, \Omega)$ is as follows: The equation*

$$\frac{Q_n(\bar{\mu}, \Omega)}{A} = \frac{Q_n(\bar{\nu}, \Omega)}{B} \tag{89}$$

has at least one real root $\Omega = \Omega_0$. In that case

$$\frac{L_n(x)}{M_L} = Q_n(x, \Omega_0).$$

NECESSITY. Put $L_n(x)/M_L = Q_n(x, \Omega_0)$, where $Q_n(x, \Omega_0)$ is an extremal polynomial of the segment $(\xi_i)_0^n = (\mu_i + \Omega\nu_i)_0$ for $\Omega = \Omega_0$.

Call this segment $(\xi_i^{(0)})_0^n$. Then its norm is

$$N(\Theta_0) = \frac{L_n(\bar{\xi}^{(0)})}{M_L} = \frac{A + \Omega_0 B}{M_L};$$

but

$$Q_n(\bar{\mu}, \Omega_0) = \frac{A}{M_L}; \quad Q_n(\bar{\nu}, \Omega_0) = \frac{B}{M_L}.$$

Consequently

$$\frac{Q_n(\bar{\mu}, \Omega_0)}{A} = \frac{Q_n(\bar{\nu}, \Omega_0)}{B} \left(= \frac{1}{M_L}\right).$$

SUFFICIENCY. Let the equation (89) have the root $\Omega = \Omega_0$. Let

$$\frac{1}{M_0} = \frac{Q_n(\bar{\mu}, \Omega_0)}{A} = \frac{Q_n(\bar{\nu}, \Omega_0)}{B}$$

and consider the polynomial $M_0 Q_n(x, \Omega_0)$. This polynomial satisfies the following conditions: its deviation is M_0;

$$M_0 \cdot Q_n(\bar{\mu}, \Omega_0) = A; \quad M_0 \cdot Q_n(\bar{\nu}, \Omega_0) = B;$$

$$Q_n(\bar{\xi}^{(0)}, \Omega_0) = \frac{A + \Omega_0 B}{M_0} = N(\Omega_0).$$

But the polynomial $L_n(x)/M_L$ satisfies

$$\frac{L_n(\bar{\xi}^{(0)})}{M_L} \leqq \frac{A + \Omega_0 B}{M_0} = N(\Omega_0).$$

Consequently

$$\frac{A + \Omega_0 B}{M_L} \leqq \frac{A + \Omega_0 B}{M_0}.$$

Hence $M_L \geqq M_0$. But $M_L > M_0$ is impossible since $M_L = \min$ by hypothesis. Consequently $M_L = M_0$ and $M_0 \cdot Q_n(x, \Omega_0) = L_n(x)$.

COROLLARY 1. *Only the ratio of A and B is significant, not their individual values.*

COROLLARY 2. *If* (89) *has several real roots, then to each root* $\Omega_1, \Omega_2, \cdots$ *such that* $Q_n(x, \Omega_1) \neq Q_n(x, \Omega_2), \cdots$, *there corresponds a polynomial* $L_n^{(1)}(x)$, $L_n^{(2)}(x), \cdots$, *and in this case Pšeborskiĭ's problem has more than one solution.*

COROLLARY 3. *Every polynomial* $P_n(x)$ *is a polynomial of least deviation for a Pšeborskiĭ problem for suitable choice of the segments* $(\mu_i)_0^n$ *and* $(\nu_i)_0^n$ *(cf. Corollary* 7 *of Theorem* 55).

In fact, if M is the deviation of $P_n(x)$ on $[0,1]$, put $Q_n(x) = P_n(x)/M$, and let $(\overset{\pm}{\sigma}_i)_1^s$ be its (complete) distribution. Construct an arbitrary segment $(\xi_i)_0^n$ with this distribution. We write ξ_i in any way as the sum of two terms: $\xi_i = \mu_i + \nu_i\lambda$; let $P_n(\bar{\mu}) = A$ and $P_n(\bar{\nu}) = B$. Then we have

$$\frac{1}{M} = \frac{Q_n(\bar{\mu})}{A} = \frac{Q_n(\bar{\nu})}{B}.$$

Thus the hypotheses of Theorem 56 are satisfied and $P_n(x)$ is a polynomial of least deviation on $[0,1]$ among those whose coefficients satisfy $\sum_0^n p_i\mu_i = A$ and $\sum_0^n p_i\nu_i = B$.

REMARK. Theorem 56 cannot be applied if one of A and B is zero. In that case let

$$\sum_0^n p_i\mu_i = A \ \ (\neq 0); \quad \sum_0^n p_i\nu_i = 0;$$

putting $\lambda_i = \mu_i + \nu_i$, we replace these two conditions by the equivalent conditions $\sum p_i\mu_i = A$ and $\sum p_i\lambda_i = A$; then Theorem 56 is applicable.

We shall give some examples of the application of Theorem 56. We begin with Zolotarev's problem.

EXAMPLE 46 (ZOLOTAREV). The conditions on the coefficients of the required polynomial reduce to the two equations $p_{n-1} = -\sigma$ $(\sigma > 0)$ and $p_n = 1$. We are to find the polynomial of least deviation on $[0,1]$.

Taking account of Corollary 1 of Theorem 56 and the fact that σ is arbitrary, we can replace these equations by $-p_{n-1} = +1$ and $p_n = \bar{\sigma}$. Then

$$(\mu_i)_0^n = 0_0, \cdots, 0_{n-2}, -1_{n-1}, 0; \quad (\nu_i)_0^n = 0_0, \cdots; 0_{n-1}, 1; \ A = 1; \ B = \bar{\sigma};$$
$$(\mu_i + \Omega\nu_i)_0^n = 0_0, \cdots, 0_{n-2}, -1_{n-1}, \Theta$$

(if we write $\Omega\bar{\sigma} = \Theta$). We obtain the segment investigated in Examples 29 and 34. For any Θ in $(-\infty, \infty)$ this segment is of class II. Let a principal polynomial for it be

$$Q_n(x, \Theta) = \sum_0^n q_i(\Theta) x^i.$$

The deformation parameter Θ yields a polynomial of the constant passport $[n, n, 0]$ in the critical interval $-\frac{1}{2}(n+1) < \Theta < -\frac{1}{2}(n-1)$. At the left-hand and right-hand endpoints of this interval, $Q_n(x,\Theta)$ is deformed into $-T_n(x)$ and $+T_n(x)$, respectively.

At the focus $\Theta = -n/2$ and $Q_n(x, -n/2) = -T_{n-1}(x)$.

Equation (89) is transformed into

$$-\frac{q_{n-1}}{1} = \frac{q_n(\Theta)}{\sigma} \quad \text{or} \quad q_n(\Theta) = -\bar{\sigma} \cdot q_{n-1}(\Theta).$$

To establish the existence of the root Θ_0 for each $\bar{\sigma}$, $-\infty < \bar{\sigma} < +\infty$, we consider the intervals in which $q_n(\Theta)$ and $q_{n-1}(\Theta)$ vary. It is easy to see that Θ varies over $-\infty < \Theta < \infty$ for the indicated $\bar{\sigma}$; then $-2^{2n-1} \leqq q_n(\Theta) \leqq 2^{2n-1}$ and $-n2^{2n-2} \leqq q_{n-1}(\Theta) \leqq n2^{2n-2}$.

The curves $y = q_n(\Theta)$ and $y = -\bar{\sigma}q_{n-1}(\Theta)$ clearly intersect just once for each $\bar{\sigma}$, and the hypothesis of Theorem 56 is satisfied. Hence this problem is equivalent to finding an extremal polynomial $Q_n(x, \Theta_0)$ for the segment $0_0, 0_1, \cdots, 0_{n-2}, -1_{n-1}, \Theta$, i.e. for the corresponding Markov problem.

EXAMPLE 47 (ZOLOTAREV). In the set of polynomials $P_n(x) = \sum_0^n p_i x^i$ that satisfy $P_n(\rho) = A$ and $p_n = 1$, find the one that deviates least from zero on $[0, 1]$ ($\rho > 1$; $A > 0$; $B = 1$).

Here $(\mu_i)_0^n = 1, \rho, \rho^2, \cdots, \rho^n$; $(\nu_i)_0^n = 0_0, 0_1, \cdots, 0_{n-1}, 1$;

$$(\mu_i + \Theta\nu_i)_0^n = 1, \rho, \rho^2, \cdots, \rho^{n-1}, \rho^n + \Theta.$$

This segment was studied in Examples 32 and 33.

Here equation (89) takes the form

$$Q_n(\rho, \Theta) = Aq_n(\Theta),$$

where $Q_n(x, \Theta) = \sum_0^n q_k(\Theta) x^k$ is an extremal polynomial for the segment $(\mu_i + \Theta\nu_i)_0^n$.

In the present case we have

$$-T_n(\rho) \leqq Q_n(\rho, \Theta) \leqq T_n(\rho) \quad \text{and} \quad -A2^{2n-1} \leqq Aq_n(\Theta) \leqq A2^{2n-1}$$

for $-\infty < \Theta < +\infty$.

Therefore the curves $y = Aq_n(\Theta)$ and $y = Q_n(\rho, \Theta)$ always intersect, and equation (89) has a real root Θ_0. Thus the Zolotarev problem with the two conditions $p_n = 1$ and $P_n(\rho) = A$ is equivalent to the problem of finding among the extremal polynomials $Q_n(x, \Theta)$ of the segment $1, \rho, \cdots, \rho^{n-1}, \rho^n + \Theta$ the polynomial $Q_n(x, \Theta_0) = \sum_0^n q_k(\Theta_0) x^k$ for which $Q_n(\rho, \Theta_0) = Aq_n(\Theta)$; and then the deviation of the required polynomial is $M = 1/q_n(\Theta_0)$ and the polynomial itself is

$$L_n(x) = \frac{Q_n(x, \Theta_0)}{q_n(\Theta_0)}.$$

EXAMPLE 48. Among the polynomials $\{P_n(x)\}$ satisfying $p_n = 1$, $p_{n-1} = 0$, find one deviating least from zero on $[0, 1]$.

We replace the given conditions by $p_n = 1$, $p_{n-1} + p_n = 1$. Then

$$(\mu_i)_0^n = 0_0, 0_1, \cdots, 0_{n-2}, 1_{n-1}, 1_n;$$

$$(\nu_i)_0^n = 0_0, \cdots, 0_{n-1}, 1_n;$$

$$(\mu_i + \Theta\nu_i)_0^n = 0_0, \cdots, 0_{n-2}, 1_{n-1}, 1 + \Theta.$$

Letting $Q_n(x,\Theta)$ be an extremal polynomial of this segment, we have (89):

$$q_n(\Theta) + q_{n-1}(\Theta) = q_n(\Theta),$$

i.e. $q_{n-1} = 0$. But in the family of polynomials $Q_n(x,\Theta)$ we have $-n2^{2n-1} \leqq q_{n-1}(\Theta) \leqq n2^{2n-1}$ for $-\infty < \Theta < \infty$. Consequently there must be a value $\Theta = \Theta_0$ for which $q_{n-1}(\Theta) = 0$. Thus the original two equations are equivalent to the single equation

$$p_{n-1} + p_n \cdot (1 + \Theta_0) = \frac{q_{n-1}(\Theta_0)}{q_n(\Theta_0)} + \Theta_0 + 1.$$

We now give an example where (89) is not satisfied.

EXAMPLE 49. Among the polynomials $\{P_n(x)\}$ satisfying $P_n(\tau_k) = A$ and $p_n = 1$, where $\tau_k = \sin^2(k\pi/n)$, find one deviating least from zero on $[0,1]$.

Let

$$(\mu_i)_0^n = 1, \tau_k, \cdots, \tau_k^n;$$

$$(\nu_i)_0^n = 0_0, \cdots, 0_{n-1}, 1;$$

$$(\mu_i + \Theta\nu_i)_0^n = 1, \tau_k, \cdots, \tau_k^{n-1}, \tau_k^n + \Theta.$$

This segment has passport $[n, n, 1(k)]$ for $k = 2, 3, \cdots, n-2$ (Chapter 5, §1).

If k is fixed and we denote this polynomial by

$$W_n(x,\Theta) = \sum_0^n w_i(\Theta)\, x^i,$$

the required equation (89) is

$$W_n(\tau_k, \Theta) = A w_n(\Theta).$$

But for $0 < \Theta < +\infty$ we have (p. 105)

$$-1 \leqq W_n(\tau_k,\Theta) \leqq +1 \quad \text{and} \quad 0 < K \leqq w_n(\Theta) \leqq 2^{2n-1}$$

(similarly for $\Theta < 0$).

Consequently it is easy to choose A so that (89) has no real root; it is enough to take $AK > 1$.

For such a value of A, Example 49 cannot be solved by using Theorem 56.

It is easy to generalize Theorem 56 to the case of three or more relations among the coefficients of the required polynomial if we make similar restrictions (§3).

§3. N. I. Ahiezer's problem and its solution by the functional method

In a memoir published in 1928, N. I. Ahiezer posed the following problem: Among the polynomials $\{P_n(x)\}$ with three given leading coefficients $p_n = 1$, $p_{n-1} = \alpha$, $p_{n-2} = \beta$ find the polynomial $Y_n(x, \alpha, \beta)$ deviating least from zero on $[-1, 1]$, and the number

$$L(\alpha, \beta) = \max_{[-1,+1]} |Y_n|.$$

He expressed these polynomials in terms of automorphic functions by using conformal mapping.

We shall extend the problem in the following way.

Among $\{P_n(x)\}$ whose coefficients satisfy three (consistent) linear relations

$$\sum_0^n p_i\mu_i = A; \quad \sum_0^n p_i\nu_i = B; \quad \sum_0^n p_i\lambda_i = C,$$

find the one that deviates least from zero on $[0, 1]$.

To solve this problem we generalize Theorem 56 to the case of three relations.

THEOREM 57. *If we seek a polynomial $L_n(x)$ deviating least from zero on $[0,1]$ among the polynomials $\{\sum_0^n p_i x^i\}$ whose coefficients satisfy the three relations*

$$\sum_0^n p_i\mu_i = A; \quad \sum_0^n p_i\nu_i = B; \quad \sum_0^n p_i\lambda_i = C,$$

and if the segment $(\xi_i)_0^n = (\mu_i + \Theta\nu_i + \Theta'\lambda_i)_0^n$ with parameters Θ and Θ' has $Q_n(x, \Theta, \Theta')$ for extremal polynomial, then a necessary and sufficient condition for the equation $Q_n(x, \Theta_0, \Theta_0') = L_n(x)/M_L$, where M_L is the deviation of $L_n(x)$, is that the system

$$\frac{Q_n(\bar{\mu}, \Theta, \Theta')}{A} = \frac{Q_n(\bar{\nu}, \Theta, \Theta')}{B} = \frac{Q_n(\bar{\lambda}, \Theta, \Theta')}{C} \tag{90}$$

has at least one pair of real roots $\Theta = \Theta_0$ and $\Theta' = \Theta_0'$.

The proof is similar to the previous one:

Let $L_n(x)/M_L = Q_n(x, \Theta_0, \Theta_0')$; the norm of the segment

$$(\xi_i^0)_0^n = (\mu_i + \Theta_0\nu_i + \Theta_0'\lambda_i)_0^n$$

is

$$N(\Theta_0, \Theta_0') = \frac{L_n(\overline{\xi^{(0)}})}{M_L} = \frac{A + \Theta_0 B + \Theta_0' C}{M_L};$$

but

$$Q_n(\bar{\mu}, \Theta_0, \Theta_0') = \frac{A}{M_L}; \quad Q_n(\bar{\nu}, \Theta_0, \Theta_0') = \frac{B}{M_L}; \quad Q_n(\bar{\lambda}, \Theta_0, \Theta_0') = \frac{C}{M_L};$$

consequently we have

$$\frac{Q_n(\bar{\mu}, \Theta_0, \Theta_0')}{A} = \frac{Q_n(\bar{\nu}, \Theta_0, \Theta_0')}{B} = \frac{Q_n(\bar{\lambda}, \Theta_0, \Theta_0')}{C} \left(= \frac{1}{M_L} \right).$$

Let equation (90) have roots $\Theta = \Theta_0$ and $\Theta' = \Theta_0'$. Let $1/M_0$ be the common value of the equal ratios in (90) and consider $M_0 \cdot Q_n(x, \Theta_0, \Theta_0')$:

$$M_0 Q_n(\bar{\mu}, \Theta_0, \Theta_0') = A; \quad M_0 Q_n(\bar{\nu}, \Theta_0, \Theta_0') = B; \quad M_0 Q_n(\bar{\lambda}, \Theta_0, \Theta_0') = C;$$

$$Q_n(\overline{\xi^{(0)}}, \Theta_0, \Theta_0') = \frac{A + \Theta_0 B + \Theta_0' C}{M_0} = N(\Theta_0, \Theta_0');$$

$$\frac{L_n(\overline{\xi^{(0)}})}{M_L} \leqq \frac{A + \Theta_0 B + \Theta_0' C}{M_0};$$

consequently

$$\frac{A + \Theta_0 B + \Theta_0' C}{M_L} \leqq \frac{A + \Theta_0 B + \Theta_0' C}{M_0}$$

and $M_L \geqq M_0$. Since $M_L > M_0$ is impossible, we have $M_L = M_0$ and $M_0 Q_n(x, \Theta_0, \Theta_0') = L_n(x)$.

EXAMPLE 50. We show that Ahiezer's problem, formulated for the interval $[0,1]$, is identical with the problem of finding an extremal polynomial for the segment-functional $0_0, 0_1, \cdots, 0_{n-3}, 1, \Theta_1, \Theta_2$, i.e. with the construction of all polynomials of passport $[n, n-1, 0]$ (and higher passports).

In fact, retaining the notation of Theorem 57, we replace the original hypotheses by equivalent ones,

$$p_n = \gamma; \quad p_{n-1} = \delta; \quad p_{n-2} = 1;$$

then

$$(\mu_i)_0^n = 0_0, 0_1, \cdots, 0_{n-3}, 1, 0, 0; \quad (\nu_i)_0^n = 0_0, 0_1, \cdots, 0_{n-2}, 1, 0;$$

$$(\lambda_i)_0^n = 0_0, 0_1, \cdots, 0_{n-1}, 1 \quad \text{and} \quad (\xi_i)_0^n = 0_0, 0_1, \cdots, 0_{n-3}, 1, \Theta_1, \Theta_2.$$

We verify that the system of equations

$$\frac{q_{n-2}(\Theta_1, \Theta_2)}{1} = \frac{q_{n-1}(\Theta_1, \Theta_2)}{\delta} = \frac{q_n(\Theta_1, \Theta_2)}{\gamma}$$

has at least one pair of real zeros $\Theta_1 = \Theta_1^{(0)}$, $\Theta_2 = \Theta_2^{(0)}$; but by Theorem 53 we can replace the parameters Θ_1, Θ_2 in a one-to-one way by the leading coefficients $q_{n-1} = \vartheta_1$ and $q_n = \vartheta_2$. We have $q_{n-2}(\Theta_1, \Theta_2) = y_{n-2}(\vartheta_1, \vartheta_2)$ and the system takes the form

$$y_{n-2}(\vartheta_1, \vartheta_2) = \frac{1}{\gamma}\,\vartheta_2 \quad \text{and} \quad \vartheta_1 = \frac{\delta}{\gamma}\,\vartheta_2,$$

where

$$|y_{n-2}| \leqq \frac{n(2n-3)}{2}\,2^{2n-4}; \quad -2^{2n-1} \leqq \vartheta_2 \leqq 2^{2n-1}; \quad -n2^{2n-2} \leqq \vartheta_1 \leqq n2^{2n-2},$$

and then

$$y_{n-2}\left(\frac{\delta}{\gamma}\,\vartheta_2, \vartheta_1\right) = \frac{1}{\gamma}\,\vartheta_2;$$

under our hypotheses the continuous curve $y_{n-2}((\delta/\gamma)\,\vartheta_2, \vartheta_1)$ passes through the points

$$\vartheta_2 = +2^{2n-1}, y_{n-2} = +\frac{n(2n-3)}{2}\,2^{2n-4} \quad \text{and} \quad -2^{2n-1}, -\frac{n(2n-3)}{2}\,2^{2n-4}$$

and consequently always intersects the lines $y = (\pm 1/\gamma)\,\vartheta_2$. Hence Theorem 57 can be applied and the problem can be solved by determining $\vartheta_2^{(0)}$, $\vartheta_1^{(0)}$ with the aid of the polynomial $Q_n(x, \vartheta_1^{(0)}, \vartheta_2^{(0)})$; in fact the required $Y_n(x)$ is given by

$$Y_n(x) = \frac{\gamma}{\vartheta_2^{(0)}}\,Q_n(x, \vartheta_1^{(0)}, \vartheta_2^{(0)}),$$

where $|\gamma|/\vartheta_2^{(0)}$ is the least deviation.

CHAPTER II

BEST UNIFORM APPROXIMATION OF POLYNOMIALS AND ANALYTIC FUNCTIONS

§1. Approximation of polynomials [14]

In problems on the uniform best approximation of a given continuous function $f(x)$ on a closed interval by algebraic polynomials $P_n(x)$ an essential role is played by Čebyšev's theorem, which reads as follows. In order that the polynomial $P_n(x) = \sum_k^n a_k x^k$ give the best approximation to $f(x)$ on $[a,b]$, i.e. that $\max|f(x) - P_n(x)| = L$ be a minimum, it is necessary and sufficient that from the set of points on $[a,b]$ at which $f(x) - P_n(x) = \pm L$ (points of deviation) we can choose *at least* $n+2$ points $\sigma_1 < \sigma_2 < \cdots < \sigma_s$ $(s \geqq n+2)$ at which $f(\sigma_i - P_n(\sigma_i) = \pm L$ with alternating signs. For short this will be called the condition of "Čebyšev alternation."

A solution of the problem always exists, and the polynomial of best approximation is unique.

We introduce some definitions.

Suppose that functions $\phi_0(x), \phi_1(x), \phi_2(x), \cdots, \phi_n(x)$, continuous on $[a,b]$, are given. An expression $\sum_{k=0}^n a_k \phi_k(x)$ is called a generalized polynomial of the system.

DEFINITION. A system $[\phi_k(x)]_{k=0}^n$ of functions is called a Čebyšev system on $[a,b]$ (or for short a T-system) if each generalized polynomial in the system, not identically zero, has at most n zeros on $[a,b]$.

We remark that this condition is equivalent to the linear independence of the functions $[\phi_k(x)]$ on $[a,b]$.

The simplest examples of T-systems are

$$1, x, x^2, \cdots, x^n \text{ on an arbitrary } [a,b];$$

$$1, \cos\phi, \cos 2\phi, \cdots, \cos n\phi \text{ on } [0,\pi].$$

Let E be any closed subset of $[a,b]$.

ČEBYŠEV'S THEOREM (EXTENDED FORM) [15]. *If $f(x)$ is a given continuous function and $[\phi_k(x)]_0^n$ is a T-system on $[a,b]$, a necessary and sufficient condition that the polynomial $P_n(x) = \sum_{k=0}^n a_k\phi_k(x)$ deviates least from $f(x)$ on E is as follows: L is the deviation of $f(x)$ from $P_n(x)$, i.e.*

the maximum of $|f(x) - P_n(x)|$ *is attained on* E *at at least* $n+2$ *points of* E; *at these points* $f(x) - P_n(x)$ *takes the values* $\pm L$ *with alternating signs (Čebyšev alternation). The problem of finding* $P_n(x)$ *for a given* $f(x)$ *always has a solution, but it is not necessarily unique in the general case.*

We remark that in general the number of points $\xi \in E$ at which $f(x) - P_n(x) = \pm L$ can considerably exceed $n+2$, and can even be infinite. What is important is that we can select from them a distribution $(\overset{\pm}{\xi})_1^{n+2}$ which contains the entire interval of alternation $(q = n+1)$.

Let us consider a special case of the Čebyšev problem. Let the given function $f(x) = P_n(x)$. We are required to find $P_m(x)$ with $m < n$ so that $\max|P_n(x) - P_m(x)|$ is a minimum on $[0,1]$.

According to Čebyšev's theorem, the number s of points of deviation must exceed by at least unity the number of parameters in the required $P_m(x) = \sum_0^m a_k x^k$, i.e. $s \geqq m+2$; also we must have Čebyšev alternation. It is clear that in this case the number s of points of deviation is in general finite.

Putting $P_n(x) - P_m(x) = Y_n(x)$, we may formulate the problem in another way:

In the set of polynomials of degree n, with the leading coefficients $b_n, b_{n-1}, \cdots, b_{m+1}$ assigned, find one that deviates least from zero on $[0,1]$, and also the deviation L.

This problem can be solved completely if $Y_n(x)/L$ belongs to the system $\{Q_n(x)\}$, the set of polynomials of the second class, i.e. if $s > \frac{1}{2}n + 1$. It is sufficient for this that $m + 2 > \frac{1}{2}n + 1$. Thus $m < n < 2m + 2$.

The results obtained in Part 1 allow us to consider the polynomials $\{Q_n(x)\}$ as known.

The lowest possible passport for the required $Y_n(x)/L$ is $[n, m+2, 0]$, since $q = m + 1$ and $p + q = s - 1$. We have shown that a determining basis functional for the polynomials of passport $[n, s, 0]$ can be taken in the following form:

$$0_0, 0_1, \cdots, 0_{s-2}, 1_{s-1}, \theta_1, \theta_2, \cdots, \theta_l \ (*),$$

where $s + l = n + 1$.

In a well-defined bounded l-dimensional region $\{M\}_l$ in the domain of the independent parameters the segment (*) determines uniquely all the polynomials of passport $[n, s, 0]$. On the boundary of $\{M\}$ the number s_1 of nodes of the determining polynomial is necessarily increased: $s_1 > s$, but the number $q = s - 1$ of intervals of alternation clearly cannot be decreased by such a change. It remains to consider points outside $\{M\}_l$.

THEOREM 58. *Every polynomial of passport* $[n, s_1, q_1]$ *with* $s_1 > s$ *and* $q_1 \geqq s - 1$ *serves the segment-functional* (*) *at at least one point of* l-*dimensional space.*

In fact, from the set of points of the distribution $(\overset{\pm}{\sigma}_i)_1^{s_1}$ of the polynomial $L_n(x)$ select s points which have alternating signs. This is always possible under the hypotheses of the theorem. We obtain a subdistribution $(\overset{\pm}{\sigma}{}'_i)_1^s$. We decompose the basis $0_0, 0_1, \cdots, 0_{s-2}, 1_{s-1}$ of the segment (*) in terms of the nodes $(\sigma'_i)_1^s$. By the constant decomposition property of this basis the weights of the nodes $(\sigma'_i)_1^s$ have alternating signs. Having found these, we extend the basis as a segment according to the structure we have obtained, i.e. we put

$$\Theta_1^{(0)} = \sum_1^s \delta_i \sigma_i'^{\,s}, \quad \Theta_2^{(0)} = \sum_1^s \delta_i \sigma_i'^{\,s+1}; \cdots; \Theta_l^{(0)} = \sum \delta_i \sigma_i'^{\,s+l-1}.$$

We obtain particular values of the functional (*) which, by the fundamental criterion, serve the polynomial $L_n(x)$. This completes the proof.

It is clear that the points $(\Theta_i^{(0)})_1^l$ do not belong to the open region $\{M\}_l$.

Let $\{Q_n(x, \Theta_1, \Theta_2, \cdots, \Theta_l)\}$ denote the set of extremal polynomials that serve the functional (*) for arbitrary (finite) values $(\Theta_i)_1^l$.

Therefore every solution $Y_n(x)/L$ of our Čebyšev problem belongs to this set:

$$\frac{Y_n(x)}{L} \in \{Q_n(x, \Theta_1, \Theta_2, \cdots, \Theta_l)\}.$$

Suppose that in the family $\{Q_n(x, \Theta_1, \Theta_2, \cdots, \Theta_l)\}$ the elements of a determining functional $(\Theta_i)_1^l$ can be replaced by an equivalent system of elements, namely $(\vartheta_i)_1^l$, the leading coefficients of the polynomial (Part 1, p. 130), i.e.

$$Q_n(x, \vartheta_1, \cdots, \vartheta_l) = \vartheta_l x^n + \vartheta_{l-1} x^{n-1} + \cdots + \vartheta_1 x^{n-l+1} + y_{n-l}(\vartheta_1, \cdots, \vartheta_l)\, x^{n-l} + \cdots,$$

where (Θ_i) are connected with (ϑ_i) by a well-defined one-to-one functional relation, and the remaining coefficients $y_i(\vartheta_1, \vartheta_2, \cdots, \vartheta_l)$ are determined from a system of differential equations. Then the whole unbounded l-dimensional space of the variables $(\Theta_i)_1^l$ is mapped on the bounded l-dimensional domain of $(\vartheta_i)_1^l$, the variable coefficients of reduced polynomials for which $|\vartheta_i| \leqq |t_i|$ (the coefficients of the Čebyšev polynomial).

Supposing that the functions $y_i(\vartheta_1, \vartheta_2, \cdots, \vartheta_l)$ are known, we can proceed to solve our problem, reformulated so that it agrees notationally with the determining functional (*).

In $Y_n(x)$ let the leading $l+1$ coefficients be $b_n, b_{n-1}, \cdots, b_{n-l}$. Then the number s of nodes of $Q_n(x) = Y_n(x)/L$ and the number q of alternations satisfy $s \geqq n+1-l$ and $q \geqq n-l$. The polynomial $Q_n(x, \vartheta_1, \vartheta_2, \cdots, \vartheta_l)$ serves the functional (*) inside $\{M\}_l$ when $s = n+1-l$ and outside or on the boundary when $s > n+1-l$.

THEOREM 59. *Among the extremal polynomials* $\{Q_n(x, \vartheta_1, \cdots, \vartheta_l)\}$ *of the functional* (*), *there is a (unique) polynomial* $Q_n(x) = \sum_0^n q_i x^i$ *whose coefficients* $q_n, q_{n-1}, \cdots, q_{n-l}$ *are proportional to given numbers* $b_n, \cdots, b_{n-l}$; *and if we put* $q_k/b_k = 1/M$ *then* $|M| = L$ *and* $Y_n(x) = |M| Q_n(x)$.

In fact, we have already shown that $Y_n(x)/L \in \{Q_n(x, \vartheta_1, \cdots, \vartheta_l)\}$ for arbitrary (real) values of $b_n, b_{n-1}, \cdots, b_{n-l}$, since the given Čebyšev problem always has a solution. Consequently, we can always find $Q_n(x) = \sum_0^n q_i x^i$ in this set with

$$\frac{q_n}{b_n} = \frac{q_{n-1}}{b_{n-1}} = \cdots = \frac{q_{n-l}}{b_{n-l}}. \tag{92}$$

Then the necessary and sufficient conditions of Theorems 56 and 57 are satisfied for the case of $n+1-l$ relations, and consequently the constant of proportionality $1/M$ yields $|M| = L$ and $Y_n(x) = |M| Q_n(x)$. This completes the proof. It remains to give a process for finding $Q_n(x)$ and M.

In the family of polynomials $Q_n(x, \vartheta_1, \vartheta_2, \cdots, \vartheta_l)$ put

$$\vartheta_i = \lambda b_i \quad (i = n, n-1, \cdots, n-l+1)$$

and

$$y_{n-l}(\lambda b_{n-l+1}, \cdots, \lambda b_n) = \lambda b_{n-l}. \tag{93}$$

Equation (93) defines a unique value $\lambda = \lambda_0$ with $\lambda_0 = \max$ (according to the theorem). If $\lambda_0 < 0$, replace $Q_n(x)$ by $-Q_n(x)$. The deviation is $L = 1/\lambda_0$.

§2. Approximation of analytic functions [16]

Let $f(x) = \sum_0^\infty \alpha_i x^i$ for each point of $[0,1]$. We are to find $Y_h(x)$ of degree at most h so that $\max_{[0,1]} |f(x) - Y_h(x)| = L(\min)$.

Let the points of deviation (taking account of sign) form the distribution $(\overset{\pm}{\sigma}_i)_1^s$; call this the f_h-distribution. It is clear that for suitable f this distribution can be any prescribed distribution as long as it is finite and has Čebyšev alternation. It is equally clear that we can restrict ourselves to functions of the form $f(x) = \sum_{h+1}^\infty \alpha_i x^i$. In what follows we shall suppose that the extremal polynomials $\{Q_n(x)\}$ of functionals of class II are known.

Any selection from a distribution $(\overset{\pm}{\sigma}_i)_1^s$, with the signs attached to the σ preserved, will be called a "subdistribution" of the given distribution.

Let $Q_n(x)$ of class II have $(\overset{\pm}{\sigma}_i)_1^s$ as its (complete) distribution. Then letting $(\overset{\pm}{\sigma}{}_i')_1^{\bar{s}}$ be a subdistribution of the given one, we introduce

$$R_{\bar{s}}(x) = \prod_1^{\bar{s}} (x - \overset{\pm}{\sigma}{}_i'),$$

the resolvent of the subdistribution. We have

$$R_{\bar{s}_1}(x) = \prod (x - \overset{+}{\sigma}{}_i'); \quad R_{\bar{s}_2}(x) = \prod (x - \overset{-}{\sigma}{}_i'); \quad \bar{s}_1 + \bar{s}_2 = \bar{s}.$$

Then we have (16)

$$Q_n(x) = 1 - R_{\bar{s}_1}^2(x) \cdot \phi(x) = -1 + R_{\bar{s}_2}^2(x) \cdot \psi(x),$$

where $\phi(x) \geqq 0$ and $\psi(x) \geqq 0$ on $0 \leqq x \leqq 1$ for any subdistribution of $Q_n(x)$. We recall (Chapter I, §1) that every extremal polynomial of higher degree or any (reduced) continuous extremal function with the same f_h-distribution must have the form

$$1 - R_{\bar{s}_1}^2(x)\,[\phi(x) - \lambda(x) R_{\bar{s}_2}^2(x)] = -1 + R_{\bar{s}_2}^2(x)\,[\psi(x) + R_{\bar{s}_1}^2(x)\lambda(x)],$$

which yields for $\lambda(x)$

$$-\frac{\psi(x)}{R_{\bar{s}_1}^2(x)} \leqq \lambda(x) \leqq \frac{\phi(x)}{R_{\bar{s}_2}^2(x)}.$$

The strip

$$\left[\frac{-\psi(x)}{R_{\bar{s}_1}^2(x)}, \frac{\phi(x)}{R_{\bar{s}_2}^2(x)}\right] \tag{94}$$

of width $2/R_{\bar{s}}^2(x)$ is the "strip of reducibility."

THEOREM 60. 1. *Every extremal polynomial $Q_p(x)$ of class* I *or* II *whose distribution contains as a subdistribution the f_h-distribution $(\overset{\pm}{\sigma}_i)_1^{\bar{s}}$ of a given $f(x)$ yields the unique decomposition*

$$f(x) = Y_h(x) + LQ_p(x) + LR_{\bar{s}}^2(x)\,\Omega_Q(x) \tag{95}$$

for the function, where $R_{\bar{s}}^2(x)$ is the squared resolvent of the f_h-distribution (with the corresponding variants), and $\Omega_Q(x)$ is an analytic function (clearly $h < p$).

2. *If a given $f(x)$ has the representation*

$$f(x) = Z_h(x) + MQ_p(x) + MR_{\bar{s}}^2(x)\,\Omega(x) \quad (h < p), \tag{96}$$

where $Q_p(x)$ is a (reduced) polynomial of class I *or* II, *whose distribution contains a subdistribution that has at least $h+1$ changes of sign, $R_{\bar{s}}^2(x)$ is the squared resolvent of this subdistribution, and $\Omega(x)$ is an analytic function whose values are not outside the interval* (94) *when $0 \leqq x \leqq 1$, then $Z_h(x)$ is a polynomial of least deviation from $f(x)$, and the deviation is $L = M$.*

Before giving the proof we make some remarks.

Let $(\sigma_i)_1^s$ be any s points of $[0, 1]$. The remainder on dividing $f(x)$ by a polynomial of the form $R_s^2(x) = \prod_1^s (x - \sigma_i)^2$ is the polynomial $r(x)$ of degree less than $2s$ obtained by the following process: expand $f(x)$ by Taylor's formula about σ_1,

$$f(x) = f(\sigma_1) + f'(\sigma_1)(x - \sigma_1) + (x - \sigma_1)^2 f_1(x);$$

then expand the partial quotient $f_1(x)$ about σ_2, etc., so that

$$f(x) = f(\sigma_1) + f'(\sigma_1)(x - \sigma_1) + f_1(\sigma_2)(x - \sigma_1)^2 + f_1'(\sigma_2)(x - \sigma_1)^2(x - \sigma_2) + (x - \sigma_1)^2(x - \sigma_2)^2 f_2(x) = \cdots,$$

and finally

$$f(x) = r(x) + R_s^2(x)\Omega(x). \tag{97}$$

To insure uniqueness in passing from one Taylor expansion to another it is enough to suppose, for example, that $f(x)$ is analytic inside two disks C_0 and C_1 of radius greater than 1 with centers at 0 and 1, since then the partial quotient $\Omega(x)$ is analytic in the same region.

It is easy to see that (97) is unchanged if the order of the decomposition with respect to the points is changed, and in general it is unique for a given $R_s^2(x)$. We say that $f(x)$ is divisible by $R_s^2(x)$ if and only if $r(x) \equiv 0$. Thus we have a necessary and sufficient condition for divisibility: each σ_i is a zero of $f(x)$ of multiplicity at least two.

We return to part 1 of the theorem. By hypothesis,

$$\max_{[0,1]} \frac{|f(x) - Y_h(x)|}{L} = 1 \tag{98}$$

and it is assumed at the points of f_h: $(\overset{\pm}{\sigma}_i)_1^{\bar{s}}$, and only at these points. Then with

$$F(x) = \frac{f(x) - Y_h(x)}{L} - Q(x)$$

we have $F(\sigma_i) = 0$ and $F'(\sigma_i) = 0$, the latter for points inside $[0, 1]$. Thus $f(x) - Y_h(x) - LQ(x)$ is divisible by $R_{\bar{s}}^2(x)$.

Conclusion 1 is thus established, and $\Omega(x)$ remains in the interval (94) for $0 \leqq x \leqq 1$, as follows from (95) and the structure of $Q(x)$ given in (16).

Conversely, if (96) holds, then $f(x) - Z_h(x)$ takes the values $\pm M$ at the points $(\overset{\pm}{\sigma}_i)_1^{\bar{s}}$, we have $\bar{s} \geqq h + 2$, and we have a sufficient number of alternations. In addition,

$$\max \frac{|f(x) - Z_h(x)|}{M} = 1,$$

and conclusion 2 holds.

THEOREM 61. *If* $f(x) = \sum_{h+1}^{\infty} \alpha_i x^i$ *is given,* $Y_h(x)$ *is the polynomial best approximating it on* $[0,1]$ *with deviation* L, *and* $(\overset{\pm}{\sigma_i})_1^{\bar{s}}$ *is the* f_h-*distribution, then among the extremal polynomials of the* f_h-*distribution there is always one of the form*

$$Q_{n+2\bar{s}}(x) = \sum_0^{N+2\bar{s}} \bar{q}_i x^i,$$

with sufficiently large N, *for which* $L\bar{q}_i = \alpha_i$ $(i = h+1, \cdots, N)$.

In fact, let $Q_n(x)$ be a principal polynomial of $(\overset{\pm}{\sigma_i})_1^{\bar{s}}$, i.e. one of least degree.

We have

$$Q_n(x) = 1 - R_{\bar{s}1}^2(x)\phi(x) = -1 + R_{\bar{s}2}^2(x) \cdot \psi(x)$$

with $\phi(x) \geqq 0$ and $\psi(x) \geqq 0$ on $0 \leqq x \leqq 1$. By (95),

$$f(x) = Y_h(x) + LQ_n(x) + R_{\bar{s}}^2(x) L\Omega_n(x).$$

Then the polynomial

$$L\Omega_n(x) = \sum_0^{\infty} b_i x^i = \sum_0^{N} b_i x^i + \rho_N(x)$$

satisfies

$$-\frac{L\psi(x)}{R_{\bar{s}1}^2(x)} < L\Omega_n(x) < \frac{L\phi(x)}{R_{\bar{s}2}^2(x)}; \quad 0 \leqq x \leqq 1, \tag{99}$$

and lies in the indicated strip of width $2L/R_{\bar{s}}^2(x)$ because of the absence of other points of deviation. We take $n \geqq N$ so large that $S_n(x) = \sum_0^n b_i x^i$ also lies in the strip (99).

Then

$$\begin{aligned} f(x) &= Y_h(x) + L\left[Q_n(x) + \frac{1}{L} S_n(x) \right] + R_{\bar{s}}^2(x) \rho_N(x) \\ &= Y_h(x) + LQ_{N+2\bar{s}}(x) + R_{\bar{s}}^2(x) \rho_N(x), \end{aligned}$$

and the proof is complete.

COROLLARY. $Y_h(x) = -L\sum_0^h \bar{q}_i x^i$.

We remark that when $n = N$ the polynomial $Q_{N+2\bar{s}}(x)$ is uniquely determined and in general belongs to class I. To make it clear that the property of proportionality of the coefficients that we have established is also valid for polynomials of class II of the f_h-distribution, we need a slight generalization of the properties of Cebyšev systems.

Let $f(x)$ and $\phi(x)$ be continuous and $\phi(x) \geqq 0$ on $[a,b]$; let $(\sigma_i)_1^s$ be the set of zeros of $\phi(x)$ on $[a,b]$. Let $E(\epsilon)$ be the closed set obtained from $[a,b]$ by deleting the open intervals $(\sigma_i - \epsilon, \sigma_i + \epsilon)$ and the two boundary half-intervals $[a, a+\epsilon)$; $(b-\epsilon, b]$ with a sufficiently small $\epsilon > 0$.

Consider generalized polynomials of the form

$$\phi(x)\, P_k(x,\epsilon) = \phi(x) \cdot \sum_0^k C_i x^i \quad (k = \text{const.}).$$

Then the best approximation of $f(x)$ on $E(\epsilon)$ by polynomials $\phi(x) P_k(x,\epsilon)$ satisfies Čebyšev's theorem, i.e. there is a unique polynomial $\phi(x) P_k(x,\epsilon)$ which deviates least from $f(x)$ on $E(\epsilon)$, and the points of maximum deviation of $|f(x) - P_k(x,\epsilon)|$ form a distribution $(\overset{\pm}{\xi_i})$ with $q \geqq k+1$ alternations.

The proof of the necessity and sufficiency, as well as of uniqueness, follows almost exactly the model given, for example, in S. N. Bernšteĭn's monograph [**24**]. An immediate application of Čebyšev's theorem is not possible, since the functions $(\phi(x) x^i)$ do not form a T-system.

THEOREM 62. *If*

$$f(x) = \sum_0^{N+2\bar{s}} q_i x^i = Q_{N+2\bar{s}}(x) \quad \textit{and} \quad \phi(x) = R_{\bar{s}}^2(x)\, x^{N+1}$$

(*Theorem* 61), *then for each* $k \geqq 1$ *there is a polynomial of the form*

$$R_{\bar{s}}^2(x) \cdot x^{N+1} \sum_0^k c_i x^i = R_{\bar{s}}^2(x)\, x^{N+1} \cdot P(x),$$

deviating least from $Q_{N+2\bar{s}}(x)$ *on* $[0,1]$, *and the number of points of deviation is* $s \geqq k+2$.

In fact, on $E(0,1) = E(\epsilon)$ with the points (σ) excluded there is, by p. 146), a unique polynomial of best approximation of the form $R_{\bar{s}}^2(x) x^{N+1} P_k(x,\epsilon)$. Let this polynomial be

$$Q_m(x,\epsilon) = Q_{N+2\bar{s}}(x) - R_{\bar{s}}^2(x)\, x^{N+1} P_k(x,\epsilon)$$

with deviation $L(\epsilon)$ and at least s points of deviation $(\xi_i)_1^s$. The problem is whether we can take the limit as $\epsilon \to 0$. Note that $L(\epsilon) < 1$, since

$$L(\epsilon) < \max_{[0,1]} |Q_{N+2s}(x)| = 1 \text{ for } C_i = 0.$$

By the continuity of $\sum_0^k C_i(\epsilon)x^i$ we have $S(\epsilon) \geqq k+2$ for arbitrary ϵ; $L(\epsilon)$ increases to 1 as $\epsilon \to 0$, since we have $Q_{N+2\bar{s}}(\overset{\pm}{\sigma}_i) = \pm 1$ at the points σ_i. The existence of $\lim_{\epsilon\to 0} C_i(\epsilon)$ follows from the boundedness of the coefficients. Denote the limiting polynomial by $Q_m(x)$; $\max|Q_m(x)| = 1$. The number of its nodes is $s \geqq k+2$, and the number of alternations is $q \geqq k+1$. Note that in the limit the uniqueness theorem fails.

COROLLARY. *Among the polynomials of class* II *of the f_h-distribution $(\overset{\pm}{\sigma}_i)_1^{\bar{s}}$ there are polynomials $Q_m^*(x) = \sum_0^m q_i x^i$ which for sufficiently large m have $q_i = \bar{q}_i$ $(i = 0, 1, \cdots, N)$.*

This follows immediately from the formula

$$Q_m(x) = Q_{N+2\bar{s}}(x) - R_{\bar{s}}^2(x)\, x^{N+1} \sum_0^k c_i x^i,$$

which holds for any $k > 0$ provided $k+2 > \frac{1}{2}m + 1$, which ensures that $Q_m^*(x)$ belongs to class II and yields the sufficient condition

$$m > 2(N + 2\bar{s}).$$

The theorem suggests a way of finding $Y_h(x)$ for a given $f(x) = \sum_{h+1}^{\infty} \alpha_i x^i$. If we are given a value m, we take a passport $[m, s, q]$ (q is the number of alternations of sign in the distribution) with $q \geqq h+1$, and consider the polynomials $\{Q_m(x)\}$, whose analytic form we know, of this passport. This is a family of class II and depends on l variables, where $l = m+1-s$; these can be taken to be the coefficients of x^{h+1}, $x^{h+2}, \cdots, x^{h+l}$ in $Q_m(x)$; we denote them by $\vartheta_{h+1}, \vartheta_{h+2}, \cdots, \vartheta_{h+l}$. Let $h + l = N - 1$ and put $\vartheta_i = \alpha_i / M$ $(i = h+1, \cdots, N-1)$. The remaining coefficients of $Q_m(x)$ are definite functions of (ϑ_i); put

$$q_N = q_N(\alpha_{h+1}/M, \cdots, \alpha_{N-1}/M) = \alpha_N / M;$$

then M is determined by these relations.

The polynomial $Q_m(x)$ determined in this way is the only polynomial of passport $[m, s, q]$ that can be the required polynomial $Q_m^*(x)$ (Corollary of Theorem 62). The suitability of $Q_m(x) = \sum_0^m q_i x^i$ can be verified by the criterion obtained from Theorem 60. In fact, $Q_m(x) = Q_m^*(x)$ if and only if it contains an f_h-distribution as a subdistribution, and then $M = L$, $-M\sum_0^h q_i x^i = Y_h(x)$.

Let $(\overset{\pm}{\rho}_i)_1^s$ be the (complete) distribution of $Q_m(x)$.

Then by (95) $f(x) - M\sum_{h+1}^m q_i x^i$ must have its zeros of multiplicity at least two in a subdistribution $(\overset{\pm}{\sigma}_i)_1^{\bar{s}}$ of $(\overset{\pm}{\rho}_i)_1^s$ containing at least $h+1$

variations of sign. If this requirement is satisfied and if $R_{\bar{s}}^2(x)$ is the squared resolvent of $(\bar{\bar{\sigma}}_i)_1^{\bar{s}}$ then the function $\Omega(x)$ in the identity $f(x) - M\sum_{h+1}^m q_i x^i = R_{\bar{s}}^2(x) M\Omega(x)$ cannot be outside the interval (94). Then $Q_m = Q_m^*$ and $Y_h(x)$ has been found.

If $Q_m(x)$ is not suitable then Q_m^* must first be sought among polynomials of other passports with the same m, and then with larger values of m. After a finite number of trials Q_m^*, and consequently $Y_h(x)$, will have been found.

CHAPTER III

THE FIRST DERIVATIVE FUNCTIONAL AND A. A. MARKOV'S PROBLEM

A. A. Markov [17] gave the following estimate for the derivative of an algebraic polynomial on a finite interval: if

$$\max_{[a,b]} | P_n(x) | = M,$$

then

$$\max | P_n'(x) | \leqq \frac{2n^2}{b-a} M. \tag{100}$$

The estimate (100), which reduces to equality only at the endpoints and only for $P_n(x) = T_n(x)$, is rather poor at interior points of the interval. This essential gap was filled to a certain extent by S. N. Bernšteĭn [18] who proved the inequality $(a = -1, b = +1)$

$$| P_n^{(k)}(x) | \leqq \left(\frac{k}{1-x^2}\right)^{k/2} \cdot n(n-1) \cdots (n-k+1) M. \tag{101}$$

In [17] Markov also investigated properties of the polynomials which yield the maximum modulus for the derivative at fixed interior points x of the basic interval; these investigations are incomplete and do not answer the question of the analytic properties of the maximum as a function of x.

Using the functional method, we shall give an exact majorant for the first derivative of a polynomial at the interior points of the interval, which we take to be $[0, 1]$.

§1. Extremal polynomials of the derivative functional

For the method that we use, transferring the problem to the interval $[0, 1]$ makes the calculations simpler.

We consider the derivative of $P_n(x)$ at an arbitrary point ξ as a linear functional F_ξ defined by the finite sequence $(\mu_i)_0^n$ (or segment-functional):

$$(\mu_i)_0^n = 0, 1, 2\xi, 3\xi^2, \cdots, n\xi^{n-1}. \tag{102}$$

Putting $F_\xi(x^k) = k\xi^{k-1}$ $(k = 0, 1, \cdots, n)$, we have $F_\xi(P_n) = P_n'(\xi)$;

and $N(\xi)$, the norm of the segment (102), is the exact majorant for the set $\{P_n'(\xi)\}$:

$$\max_{[0,1]}|P_n'(\xi)| \leqq M \cdot N(\xi).$$

Our problem is to find the norm and the extremal polynomials of (102). We begin with some necessary remarks.

REMARK 1. Putting $\mu_i = \mu_{i,0}$, we form the table of differences $(\mu_{k,p})$ of (102), with $\mu_{k,p} = \mu_{k,p-1} - \mu_{k+1,p-1}$; we obtain

$$(*) \qquad (\mu_{0,k})_{k=0}^n = 0, -1, -2(1-\xi), \cdots, -n(1-\xi)^{n-1}.$$

This functional has the same norm (p. 18) as (102). Consequently

$$(103) \qquad N(\xi) = N(1-\xi).$$

If an extremal polynomial of (102) is $Q_n(x)$ then $Q_n(1-x)$ is an extremal polynomial of (*) (p. 19).

REMARK 2. If $\xi < 0$ the segment (102) has alternating signs and consequently it is served by the polynomial $+T_n(x)$ for odd n, and by $-T_n(x)$ for even n. Hence for each ξ not inside $[0,1]$ the extremal polynomial is $\pm T_n(x)$.

We now find the points of $[0,1]$ where F_ξ is served by the Čebyšev polynomials $\pm T_n(x)$.

We recall the necessary and sufficient condition for a given reduced polynomial $Q_n(x)$ to serve a given segment-functional $(\mu_i)_0^n = F$.

Let $(\overset{\pm}{\sigma}_i)_1^s$ be the distribution of $Q_n(x)$, i.e. the set of its extrema on $[0,1]$ at which $Q_n(x) = \pm 1$, with a $+$ sign if $Q_n(\sigma) = +1$, and a $-$ sign if $Q_n(\sigma) = -1$.

If the system of $n+1$ equations in the s unknowns δ_i,

$$(104) \qquad \sum_{i=1}^s \delta_i \sigma_i^k = \mu_k \quad (k = 0, 1, \cdots, n),$$

satisfies the following conditions:

1) the system is consistent;
2) the system has a solution with signs such that either $\operatorname{sgn} \delta_i = Q_n(\sigma_i)$, or $\delta_i = 0$ (generally speaking, not all $\delta_i = 0$),

then these conditions are necessary and sufficient for $Q_n(x)$ to be extremal, i.e. $F(Q_n) = +N$.

Clearly when $Q_n(x) = \pm T_n(x)$, condition 1 drops out, since $s = n+1$. Note that the determinant of (104) is a Vandermonde determinant.

We now find the condition for the functional (102) to be served by the polynomials $\pm T_n(x)$.

Let $(\tau_i)_0^n$ be the extrema of T_n and let the Čebyšev resolvent be

$$R_{n+1}(x) = \prod_0^n (x - \tau_i).$$

In this case the system (104) takes the form

$$\sum_{i=0}^{n} \delta_i \tau_i^k = k\xi^{k-1} \quad (k = 0, 1, \cdots, n). \tag{105}$$

We have

$$\delta_k = \frac{(-1)^{n-k}}{\prod_{i \neq k} (\tau_k - \tau_i)} \frac{R'_{n+1}(\xi)(\xi - \tau_k) - R_{n+1(\xi)}}{(\xi - \tau_k)^2}. \tag{106}$$

This is a polynomial of degree $n - 1$ in ξ.

Let $(\vartheta_i)_1^n$ be the extrema of $R_{n+1}(\xi)$; it is clear that δ_k takes alternating signs at these points. Consequently F_{ϑ_i} is served by one of the polynomials $\pm T_n(x)$, and since $\delta_k \neq 0$ $(k = 0, 1, \cdots, n)$, the service extends to certain intervals (Chapter I). Between these, at the points $(\tau_k)_1^{n-1}$, the extremal polynomial for $F(\tau_k)$ is necessarily different, since $T'_n(\tau_k) = 0$. We shall not discuss these points at present, but they indicate that the domain where Čebyšev polynomials serve consists of at least n separate subintervals of $[0, 1]$ (we call them Čebyšev intervals). Since when $\xi = 0$ or $\xi = 1$ we have $\delta_k \neq 0$ for all k, we have $\alpha < 1$ and $\beta > 0$ for the two boundary Čebyšev intervals $(\alpha, +\infty)$ and $(-\infty, \beta)$.

THEOREM 63. *Put* $R'_{n+1}(\xi)(\xi - \tau_k) - R_{n+1}(\xi) = \Delta_k(\xi)$.

1. *Let* $R'_{n+1}(\xi) > 0$; *then if* $\Delta_0(\xi) \leqq 0$ *we have* $\Delta_k(\xi) < 0$ $(k \neq 0)$; *if* $\Delta_n(\xi) \geqq 0$ *then* $\Delta_k(\xi) > 0$ $(k \neq n)$; *if* $\Delta_k(\xi) = 0$ *then* $\Delta_{k-i}(\xi) > 0$ *and* $\Delta_{k-i}(\xi) < 0$ $(i > 0,\ k \neq 0, n)$.

2. *Let* $R'_{n+1}(\xi) < 0$; *then if* $\Delta_0(\xi) \geqq 0$ *we have* $\Delta_k(\xi) > 0$ $(k \neq 0)$; *if* $\Delta_n(\xi) \leqq 0$ *then* $\Delta_k(\xi) < 0$ $(k \neq n)$; *if* $\Delta_k(\xi) = 0$ *for* $k \neq 0, n$, *then* $\Delta_{k-i}(\xi) < 0$ *and* $\Delta_{k+i}(\xi) > 0$.

In fact, if ξ is fixed and $\Delta_k(\xi) = \phi(\tau_k)$ we have

$$\phi'(\tau) = -R'_{n+1}(\xi),$$

whence the conclusions of the theorem follow immediately.

We note that service of the segment-functional (102) by the polynomials $\pm T_n(x)$ ends (or begins) if and only if, when conditions 1 and 2 of the criterion for extremality are satisfied, at least one weight δ_i is zero (Chapter I).

COROLLARY 1. *Let* α *and* β *be the left-hand and right-hand ends of a Čebyšev interval; then at the ends of* $[\alpha, \beta]$ *one of the boundary nodes loses its weight; more specifically,* $\delta_0 = 0$ *for* $\xi = \alpha$ *and* $\delta_n = 0$ *for* $\xi = \beta$.

COROLLARY 2. α *and* β *are respectively the roots of*

$$R'_{n+1}(\alpha) \cdot \alpha - R_{n+1}(\alpha) = 0; \tag{107}$$

$$R'_{n+1}(\beta)(\beta - 1) - R_{n+1}(\beta) = 0, \tag{108}$$

with the double root $\alpha = 0$ *excluded from the first equation, and the double root* $\beta = 1$ *excluded from the second. Hence* α *and* β *satisfy equations of degree* $n - 1$; *it follows that the number of Čebyšev intervals on* $[0, 1]$, *including the boundary intervals, does not exceed* n, *and therefore is equal to* n.

Denote these intervals by

$$[0, \beta_1], [\alpha_2, \beta_2], \cdots, [\alpha_{n-1}, \beta_{n-1}], [\alpha_n, 1]$$

(we may put $\alpha_1 = 0$, $\beta_n = 1$). The alternating service of F_ξ by the polynomials $+ T_n(x)$ and $- T_n(x)$ takes place in the following way: the extremal polynomial $+ T_n(x)$ serves F_ξ on the intervals

$$[\alpha_n, 1], [\alpha_{n-2}, \beta_{n-2}], \cdots;$$

on the other intervals the extremal polynomial is $- T_n(x)$.

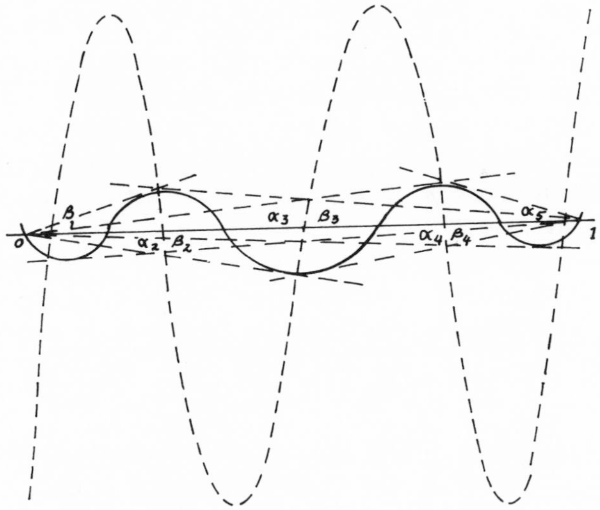

FIGURE 5.

REMARK. The nature of the distribution of the Čebyšev intervals on $[0, 1]$ is easily displayed graphically. Since $R_{n+1}(\tau_k) = 0$, we have

$R_{n+1}(x) - R_{n+1}(\tau) = (x-\tau)R_{n+1}(\xi)$, where $\tau = 0$ or 1 and $x = \alpha$ or β. But by (107) and (108) we may take $\xi = \alpha$ or β. Thus the tangents to the curve $y = R_{n+1}(x)$ drawn from the points 0 and 1 ($n-1$ of each) yield points of tangency whose abscissas are $(\alpha_i)_2^n$ and $(\beta_i)_1^{n-1}$ (Figure 5, where the construction is carried out for $n = 5$; the dotted curve is $T_5(x)$ and the solid curve is $R_6(x)$).

As we see, outside the Čebyšev intervals F_ξ is served by polynomials of passport $[n,n,0]$. These polynomials were investigated in detail in Chapter IV of the first part. We recall briefly their basic properties: they are reduced polynomials of degree n with distributions consisting of n nodes $(\overset{\pm}{\sigma}_i)_1^n$ with alternating signs; they form a family depending on a single parameter, which can be taken to be the leading coefficient. When this is done, the polynomial takes the form

$$\vartheta x^n + y_{n-1}(\vartheta)x^{n-1} + \cdots + y_1(\vartheta)x + y_0(\vartheta).$$

If $0 < \vartheta < 2^{2n-1}$, we denote the family by $Q_n(x,\vartheta)$. For $-2^{2n-1} < \vartheta < 0$ we have the polynomials $(-1)^{n-1}Q_n(1-x,\vartheta)$, and for $\vartheta = 0$ we have

$$Q_n(x,0) = -T_{n-1}(x).$$

(Note that 2^{2n-1} is the leading coefficient of $T_n(x)$, and hence it is impossible to have $|\vartheta| > 2^{2n-1}$.) The polynomials of passport $[n,n,0]$ are all of one of the forms $\pm Q_n(x,\vartheta)$ and $\pm Q_n(1-x,\vartheta)$. We call these the general Zolotarev polynomials. They fall into two classes, differing sharply in analytic (and geometric) form, as was already shown by Zolotarev. One class consists of transformations of $T_n(x)$, simply described by a magnification factor ν, so that it consists of $\pm T_n(\nu x)$ and $\pm T_n(\nu(1-x))$, where $\cos^2(\pi/2n) \leqq \nu < 1$; this guarantees the preservation of n nodes on $[0,1]$. The relation between the parameters ν and ϑ is clearly $\vartheta = 2^{2n-1}\nu^n$.

For $0 < \vartheta < 2^{2n-1}\cos^{2n}(\pi/2n)$ we have the the second class, denoted by $Z_n(x,\vartheta)$, the Zolotarev polynomials proper, which Zolotarev expressed in terms of elliptic functions. The entire set of such polynomials are expressible as

$$\pm Z_n(x,\vartheta); \quad \pm Z_n(1-x,\vartheta).$$

By the theorem on continuous deformation (Theorem 27), as ϑ decreases continuously from 2^{2n-1} to -2^{2n-1} the polynomial $Q_n(x,\vartheta)$ is deformed continuously from $+T_n(x)$ to $-T_n(x)$, following the sequence $T_n(\nu x)$, $Z_n(x,\vartheta)$, through $-T_{n-1}(x)$, then $(-1)^{n-1}Z_n(1-x,\vartheta)$, $(-1)^{n-1}T_n(\nu,(1-x))$, and finally becoming $-T_n(x)$ at $\vartheta = -2^{2n-1}$.

Note that in this process the variable nodes $\sigma_i(\vartheta)$ are displaced continuously to the right on $[0,1]$, except for the nodes 0 and 1, which are fixed for the Zolotarev polynomials proper (p. 100).

The polynomials $Z_n(x,\vartheta)$ are connected with their resolvent

$$R_n(x,\vartheta) = \prod_{i=1}^{n} (x - \sigma_i)$$

by the following fundamental relation (57):

$$\frac{\partial Z_n(x,\vartheta)}{\partial \vartheta} = R_n(x,\vartheta).$$

These facts are enough to make it possible to investigate F_ξ at the points of $[0,1]$ outside the Čebyšev intervals.

THEOREM 64. *Every polynomial $L_n(x)$ of passport $[n,n,0]$ serves F_ξ at precisely the $n-1$ points ξ at which $R'(\xi)=0$, i.e. at the extrema of its resolvent*

$$R_n(x) = \sum_{0}^{n} r_k x^k.$$

Let $(\overset{\pm}{\sigma}_i)_1^n$ be the distribution of $L_n(x)$ and let $R_n'(\xi_0)=0$. We see that $L_n(x)$ serves F_{ξ_0}. In fact, if we solve the system of n equations

$$\sum_{i=1}^{n} \delta_i \sigma_i^k = k\xi_0^{k-1} \quad (k = 0, 1, \cdots, n-1),$$

we obtain

$$\delta_k = (-1)^{n-k} \frac{R_n(\xi_0)}{\prod_{i \neq k} |\sigma_k - \sigma_i| \, (\xi_0 - \sigma_k)^2};$$

consequently condition 2 of the criterion for extremality (alternation of signs) is satisfied for all (σ_i).

Condition 1, the requirement that

$$n\xi_0^{n-1} = \sum_{1}^{n} \delta_i \sigma_i^k,$$

i.e. the condition that the whole system (105) is consistent, leads to

$$F_\xi(R) = R_n(\xi) = \sum_{k=0}^{n} r_k \sum_{i=1}^{n} \delta_i \sigma_i^k = \sum_{1}^{n} \delta_i R_n(\sigma_i),$$

i.e. $R_n'(\xi) = 0$, which is satisfied at ξ_0.

This completes the proof of Theorem 64.

We denote by (β, α) any subinterval of $[0,1]$ between two adjacent Čebyšev intervals; there are $n-1$ such subintervals. We have shown that each polynomial $L_n(x)$ serves F_ξ at exactly $n-1$ points, the roots of $R_n'(\xi) = 0$, lying one each in the intervals (β, α).

Let us suppose for definiteness that $+T_n(x)$ serves F_ξ in the first Čebyšev interval to the left of (β, α).

COROLLARY. *Since F_ξ loses its weight at $\tau_n = 1$ or $\tau_0 = 0$ at the endpoints β and α, when ξ varies from β to α within (β, α) service passes from one general Zolotarev polynomial to another in the order indicated above, and by the theorem on continuous deformation it passes through them all, ending with $-T_n(x)$ at α.*

Hence F_ξ is served only by Zolotarev polynomials on (β, α), and these are, up to sign, all the polynomials of passport $[n, n, 0]$. We call these intervals the Zolotarev intervals. In Zolotarev intervals adjacent to (β, α) the serving polynomials differ only in sign.

Thus the derivative functional (102) is stable. Other important examples of stable functionals are the trigonometric functionals

$$F_{\cos} = (\cos k\phi)_{k=0}^n \quad \text{and} \quad F_{\sin} = (\sin k\phi)_{k=0}^n$$

(Chapter IV), and also the functionals given by higher derivatives of polynomials.

§2. The norm of the derivative functional on $[0,1]$

Let E_T be the set of points belonging to Čebyšev segments, and E_Z the set of points belonging to (open) Zolotarev intervals. Then by the results of Part 1 we have

$$|P_n'(\xi)| \leq \begin{cases} |T_n'(\xi)| = N(\xi) & \text{for } \xi \in E_T, \\ |Z_n'(\xi, \vartheta_\xi)| = N(\xi) & \text{for } \xi \in E_Z \end{cases} \qquad [\max_{[0,1]} |P_n(x)| = 1].$$

In the second case ϑ_ξ is the root of $R_n'(\xi, \vartheta) = 0$.

The function $N(\xi)$ is continuous (p. 21), and by (103) it is enough to consider it on half the interval $[0,1]$.

THEOREM 65. *The zeros $(\gamma_i)_1^{n-2}$ of $T_n''(x)$ lie, one each, in the interior Čebyšev intervals.*

Let $[\alpha, \beta]$ be one of these intervals with $\alpha > \frac{1}{2}$. It follows from the equation

$$n 2^{2n-1} R_{n+1}(x) = T_n'(x) x(x-1)$$

that $R_{n+1}(x)$ and $T_n'(x)$ have the same sign on $[0,1]$.

In addition, $\operatorname{sgn} R_{n+1}(\alpha) = \operatorname{sgn} R_{n+1}(\beta)$.

Therefore

$$(109) \qquad n2^{2n-1}R_{n+1}'(x) = T_n''(x)x(x-1) + T_n'(x)(2x-1).$$

It follows from (107) and (108) that

$$n2^{2n-1}R_{n+1}(\alpha) = T_n''(\alpha)\alpha^2(\alpha-1) + T_n'(\alpha)\alpha(2\alpha-1)\ (= I_1' + I_2');$$

$$n2^{2n-1}R_{n+1}(\beta) = T_n''(\beta)\beta(\beta-1)^2 + T_n'(\beta)(\beta-1)(2\beta-1)\ (= I_1'' + I_2'').$$

If $R_{n+1}(\alpha) > 0$ and $R_{n+1}(\beta) > 0$, we have $I_1' > 0$ since $I_2' < 0$, i.e. $T_n''(\alpha) < 0$.

Moreover, $I_2'' > 0$.

We have shown that

$$n2^{2n-1}R_{n+1}(\beta) > T_n'(\beta)(2\beta-1)(\beta-1),$$

but this is equivalent to the statement that $\beta < 1$.

Thus $I_1'' > 0$ and consequently $T_n''(\beta) > 0$.

If $R_{n+1}(\alpha) < 0$ and $R_{n+1}(\beta) < 0$, the proof is similar.

THEOREM 66. *The roots $(\vartheta_i)_1^n$ of $T_n(x)$ lie, one each, in the Čebyšev intervals.*

1. Let $[\alpha,\beta]$ be an interior interval with $\alpha > \frac{1}{2}$.

With x replaced by $2x-1$ the differential equation for $T_n(x)$ is

$$(110) \qquad x(1-x)T_n''(x) - (x - \tfrac{1}{2})T_n'(x) + n^2T_n(x) = 0.$$

For definiteness, suppose that F_ξ is served by $+T_n(x)$ on $[\alpha,\beta]$ and let γ be the root of $T_n''(x)$ in this interval. Since $T_n'(x) > 0$ here, it follows from (110) that $T_n(\gamma) > 0$, and a fortiori $T_n(\beta) > 0$. It remains to verify that $T_n(\alpha) < 0$. We have $R_{n+1}'(\alpha) < 0$, and then it follows from (109) that $T''(\alpha) > 0$.

Hence it follows from (110) that

$$n^2T_n(\alpha) = (\alpha - \tfrac{1}{2})T_n'(\alpha) - \alpha(1-\alpha)T_n''(\alpha).$$

But (109) implies that

$$-T_n''(\alpha)\alpha(1-\alpha) + T_n'(\alpha)(2\alpha-1) < 0,$$

and consequently $T_n(\alpha) < 0$.

2. For the last Čebyšev interval $[\alpha,1]$ we have $T_n''(x) > 0$, $T_n'(x) > 0$. Hence it is still true here that $T_n'(\alpha) > 0$, and it follows from (109) and (110) that $T_n(\alpha) < 0$. This establishes Theorem 66.

COROLLARY OF THEOREM 65. *In each interior Čebyšev interval the norm takes its maximum* $N(\gamma_k) = |T_n'(\gamma_k)|$ *just once. In the boundary Čebyšev intervals the norm decreases monotonically from outside in, i.e.*

$$\max N(\xi) = T_n'(1) = |T_n'(0)|.$$

REMARK. If n is odd, there is a Čebyšev interval of the form $[\alpha, 1-\alpha]$ containing the point $\frac{1}{2}$. In this case $\gamma = \vartheta = \frac{1}{2}$ and Theorems 65 and 66 remain valid. In each of the other Čebyšev intervals (to the right of $\frac{1}{2}$) we have $\gamma > \vartheta$, since it follows from (110) that $\operatorname{sgn} T_n(\gamma) = \operatorname{sgn} T'(\gamma)$. Thus

$$N(\xi) = |T_n'(\xi)| = \frac{n|\sin n\Theta|}{\sqrt{\xi(1-\xi)}}$$

on $[\alpha, \beta]$, where $\Theta = \arccos(2\xi - 1)$.

We note a further corollary. The inequality

$$N(\xi) \leqq \frac{n}{\sqrt{\xi(1-\xi)}} \quad (\xi \in [\alpha, \beta]) \tag{111}$$

becomes an equality only at the points $|\sin n\Theta| = 1$, i.e. at the roots $(\vartheta_i)_1^n$ $(\vartheta_i = \cos^2((2i-1)n/4n))$ of the polynomial $\cos n\Theta = T_n(x)$. Therefore $\max N(\xi)$ increases from one interval to the next on $(\frac{1}{2}, 1)$ like the sequence

$$2n, \frac{2n}{\sin((n-2)\pi/2n)}, \cdots, \frac{2n}{\sin(\pi/2n)}, 2n^2 \quad (n \text{ odd}).$$

This also follows from (101) with $k = 1$; for a reduced polynomial $P_n(x)$ on $(-1, 1)$ this takes the form

$$|P_n'(x)| \leqq \frac{n}{\sqrt{x(1-x)}}. \tag{112}$$

The Bernšteĭn majorant (112) is thus the exact majorant of $N(\xi)$ at the points $(\vartheta_i)_1^n$.

Now consider an arbitrary Zolotarev interval. According to §1, there is a unique point ξ^* in (β, α) at which F_ξ is served by $-T_{n-1}(x)$.

Let (β, ξ^*) and (ξ^*, α) be the left-hand and right-hand parts of (β, α). The point ξ^* is defined as a root of $R_n'(\xi) = 0$, where $R_n(x)$ is the resolvent of $T_{n-1}(x)$:

$$R_n(x) = \prod_{i=0}^{n-1} (x - \tau_i'),$$

where $T_{n-1}(\tau_i') = \pm 1$.

THEOREM 67. *On* (β, α) *the norm* $N(\xi)$ *varies monotonically at each point* ξ *at which the second derivative of the extremal polynomial is not zero.*

Suppose for definiteness that $N(\beta) = + T_n'(\beta)$. Then (p. 160) in some interval (β, A), where $\beta < \xi < A < \xi^*$, the extremal polynomials are $T_n(\nu x)$, and they serve F_ξ for $\cos^2(\pi/2n) < \nu < 1$. We have

$$N(\xi) = T_n'(\nu\xi) \cdot \nu, \qquad N'(\xi) = - T_n'(\nu\xi)\frac{\beta}{\xi^2}.$$

The relation between ν and ξ is found as follows: the quantity $T_n'(\nu\xi)\cdot\nu$ must be a maximum with respect to ν, i.e. $T_n''(\nu\xi)\nu\xi + T_n'(\nu\xi) = 0$, whence it follows that $\nu\xi = \text{const.}$ (and it is greater than zero and different in different intervals). Therefore $\nu = \beta/\xi$ (when $\nu = 1$, $\xi = \beta$). We obtain $T_n'(\nu\xi) > 0$, $N'(\xi) < 0$, and $N(\xi)$ decreases. Since $\nu\xi = \beta$ for (β, α), we have $A = \beta/\cos^2(\pi/2n)$. For the other Čebyšev transformations there is a similar proof that the norm is monotonic.

Thus the extrema of $N(\xi)$ lie on the part of (β, α) where F_ξ is served by the Zolotarev polynomials proper (p. 160). It remains to establish Theorem 67 for these polynomials.

Let $A < \xi_1 < \xi_1^*$ and let an extremal polynomial for F_{ξ_1} be $Z_n(x, \vartheta_{\xi_1})$ with resolvent $R_n(x, \vartheta_{\xi_1})$. Then

$$(113) \qquad N(\xi_1) = \left(\frac{\partial Z_n(\xi, \vartheta_{\xi_1})}{\partial \xi}\right)_{\xi=\xi_1} \quad \text{and} \quad \left(\frac{\partial R_n(\xi, \vartheta_{\xi_1})}{\partial \xi}\right)_{\xi=\xi_1} = 0$$

(the last equation follows from Theorem 64). We have

$$N'(\xi) = \frac{\partial^2 Z_n(\xi, \vartheta)}{\partial \xi^2} + \frac{\partial^2 Z_n(\xi, \vartheta)}{\partial \xi\, \partial \vartheta}\frac{d\vartheta}{d\xi}$$

and by (57) and (113)

$$N'(\xi_1) = \left(\frac{\partial Z_n}{\partial \xi^2}\right)_{\xi=\xi_1} + \left(\frac{\partial R_n(\xi, \vartheta)}{\partial \xi}\frac{d\vartheta}{d\xi}\right)_{\substack{\xi=\xi_1 \\ \vartheta=\vartheta_{\xi_1}=\vartheta_1}} = Z_n''(\xi_1, \vartheta_1).$$

This completes the proof.

THEOREM 68. *In each interval* (β, α) *there is a unique point* ξ_0 *at which* $N'(\xi) = 0$, *and*

$$N(\xi_0) = \min_{(\beta,\alpha)} N(\xi) \leqq |T_{n-1}'(\xi^*)|;$$

if $\beta > \frac{1}{2}$ *then* $\beta < \xi_0 < \xi^*$; *if* $\alpha < \frac{1}{2}$ *then* $\xi^* < \xi_0 < \alpha$.

In fact, for $Z_n(x, \vartheta)$ and its resolvent $R_n(x, \vartheta)$ we have

$$Z_n'(x, \vartheta) = n\vartheta \frac{R_n(x, \vartheta)(x - \lambda)}{x(x-1)}, \tag{114}$$

where $\lambda = \lambda(\vartheta)$ is the zero of $Z_n'(x, \vartheta)$ outside $[0, 1]$. If ϑ decreases from $2^{2n-1}\cos^{2n}(\pi/2n)$ to zero, λ increases from 1 to ∞.

Let $Z_n''(\xi_0, \vartheta_{\xi_0}) = 0$. Since $Z_n'(\xi_0, \vartheta_{\xi_0}) > 0$, it follows from (114) that $R_n(\xi_0, \vartheta_{\xi_0}) > 0$. From (113) we have

$$N'(\xi_0) = Z_n''(\xi_0, \vartheta_{\xi_0}) = n\vartheta R_n(\xi_0, \vartheta_{\xi_0}) \left[\frac{\xi - \lambda}{\xi(\xi - 1)}\right]'_{\xi = \xi_0}.$$

Consequently ξ_0 can be found from the equation

$$\lambda(2\xi_0 - 1) - \xi_0^2 = 0.$$

Thus when $\alpha < \frac{1}{2}$ there is no such point ξ_0, and in the left-hand part of the interval (β, α) the necessary condition $\min N(\xi) = N(\xi_0)$ is possible only when $\beta > \frac{1}{2}$. By the symmetry of $N(\xi)$ this proves the second part of the theorem. It remains to find ξ_0 for $Z_n(x, \vartheta)$ when $\beta > \frac{1}{2}$. To do this it is convenient to replace ϑ by λ and consider $Z_n(x, \lambda)$ for $1 < \lambda < +\infty$, where $\lambda(\xi)$ increases monotonically with ξ on (A, ξ^*). Put $\lambda = \phi(\xi)$. At ξ_0 we have simultaneously $\lambda = \phi(\xi)$ and $\lambda = \xi^2/(2\xi - 1)$; since the first curve increases monotonically from 1 to ∞ on (A, ξ^*) and the second (a hyperbola) decreases monotonically from $+\infty$ to 1 on $(\frac{1}{2}, 1)$, it follows that the point of intersection is unique.

REMARK. For a Zolotarev interval of the form $(\beta, 1 - \beta)$, i.e. for even n, we have $\xi_0 = \xi^* = \frac{1}{2}$ and $\min N(\xi) = 2(n-1)$.

In comparing the appearance of the successive curves $N(\xi)$ for different n, we denote the one corresponding to $F^{(n)}(\xi)$ by $N_n(\xi)$. Note that two curves $N_n(\xi)$ and $N_{n-1}(\xi)$ cannot have more than one intersection in each Zolotarev interval for $F^{(n)}(\xi)$, since otherwise there would be a contradiction of the theorem on the uniqueness of the extremal polynomial. At a point ξ^* of tangency,

$$N(\xi^*) = |T_{n-1}'(\xi^*)| = N_{n-1}(\xi^*),$$

but

$$\min_{(\beta,\alpha)} N_n(\xi) < |T_{n-1}'(\xi^*)| \text{ for } \beta < \xi^* < \alpha.$$

In Figure 6 the solid lines are the graphs of $y = N_n(\xi)$ for $n = 2, 3, 4, 5$, and the dotted line is the Bernšteĭn majorant $y = 4/\sqrt{x(1-x)}$ (for $n = 4$). It generally gives a poor estimate for $N(\xi)$ at the points of E_Z.

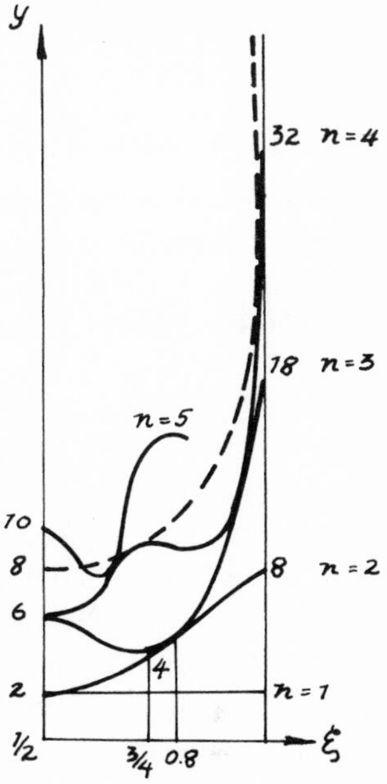

FIGURE 6.

We note that A. A. Markov's theorem was generalized by V. A. Markov [12], who proved that

$$\max_{[-1,+1]} |P_n^{(k)}(x)| \leqq MT_n^{(k)}(1).$$

But neither A. A. Markov nor V. A. Markov, in studying the question of a bound for the derivatives at interior points of the fundamental interval, took advantage of the use of the Zolotarev polynomials ([17], p. 64 and [12], p. 55), and hence they could not carry the problem to completion.

V. A. Markov's problem was completely solved in 1961 by V. A. Gusev by the functional method. We mention only one general result from his paper.

The sum of the lengths of the Čebyšev intervals on $[0, 1]$ *for the kth derivative functional on polynomials of degree n is equal to* k/n.

For the proof we refer to Gusev's paper [19].[1)]

[1)] *Editor's note*: A translation of this paper appears as an appendix to the present translation.

Chapter IV

The Trigonometric Functionals $F_{\rho\cos}$ and $F_{\rho\sin}$ and Inequalities for Polynomials in the Complex Plane

§1. Stability of trigonometric functionals [20]

The trigonometric polynomials

$$C_n(\phi) = \sum_{k=0}^{n} a_k \cos k\phi \quad \text{and} \quad S_n(\phi) = \sum_{k=1}^{n} b_k \sin k\phi$$

can be considered as finite functionals $F_{\cos} = 1, \cos\phi, \cos 2\phi, \cdots, \cos n\phi$ and $F_{\sin} = 0, \sin\phi, \sin 2\phi, \cdots, \sin n\phi$ on algebraic polynomials.

It is sufficient to consider $F_{\cos}$ for $0 < \phi \leqq \pi$ and $F_{\sin}$ for $0 < \phi < \pi$.

We have shown (Examples 30, 31) that $F_{\cos}$ and $F_{\sin}$, having fictitious double-nodal distributions, are subject to the restriction $s \geqq n$, where s is the number of faithful nodes of either one.

Similarly in the general case we have

$$F_{\rho\cos} = 1, \rho\cos\phi, \rho^2\cos 2\phi, \cdots, \rho^n \cos n\phi; \tag{115}$$

$$F_{\rho\sin} = 0, \rho\sin\phi, \rho^2\sin 2\phi, \cdots, \rho^n \sin n\phi. \tag{116}$$

$$\rho > 0$$

Since

$$\rho^k \cos k\phi = \frac{1}{2}(\rho e^{\phi i})^k + \frac{1}{2}(\rho e^{-\phi i})^k;$$

$$\rho^k \sin k\phi = \frac{1}{2i}(\rho e^{\phi i})^k - \frac{1}{2i}(\rho e^{-\phi i})^k,$$

both functionals have a fictitious double-nodal structure. Thus the functionals (115) and (116) belong to class II for $0 < \phi < \pi$ and $\rho > 0$, and at each point (ρ, ϕ) there is a unique extremal polynomial for each.

Choose a point $z = \rho e^{\phi i}$ in the upper half of the complex plane, mark the Čebyšev nodes $(\tau_k)_0^n$ on $[0,1]$, and put $\phi = \phi_0$. Then putting $\arg(z - \tau_k) = \phi_k$, we have for each $\rho = \text{const.}$

$$\phi_0 < \phi_1 < \cdots < \phi_n \quad (0 < \phi < \pi).$$

With the notation $\Psi_k = \sum_{j \neq k} \phi_j$ $(k = 0, 1, \cdots, n)$ we can write

$$\Psi_n < \Psi_{n-1} < \cdots < \Psi_0 \text{ for } 0 < \phi < \pi.$$

We note that

$$\Psi_0 - \Psi_n = \phi_n - \phi_0 < \pi. \tag{117}$$

THEOREM 69. *If $E_c(\rho)$ is the set of points of the z-plane at each of which $F_{\rho\cos}$ is served by $+T_n(x)$ or $-T_n(x)$, then a necessary and sufficient condition that a given point $\rho e^{\phi i} \in E_c$ is as follows: the $(\Psi_p)_0^n$ (after reduction) lie either in $[-\pi/2, \pi/2]$ (right-hand half-plane) or in $[\pi/2, 3\pi/2]$ (left-hand half-plane), and then $+T_n(x)$ serves in the first case, $-T_n(x)$ in the second.*

We decompose $F_{\rho\cos}$ in terms of the nodes $(\tau_k)_0^n$. For the weights $(\delta_k)_0^n$ we have

$$\delta_p = (-1)^{n-p} \frac{\frac{1}{2}\left[\prod_{k\neq p} (\rho e^{\phi i} - \tau_k) + \prod (\rho e^{-\phi i} - \tau_k)\right]}{\prod_{k \neq p} |\tau_p - \tau_k|} \qquad (p = 0, 1, \cdots, n).$$

Here the numerator $A_p = \operatorname{Re} \prod (\rho e^{\phi i} - \tau_k)$. The condition that $F_{\rho\cos}$ is served by the Čebyšev polynomial (27) is that the $(\delta_p)_0^n$ have alternating signs (or possibly are zero). This requires that either $A_p \geqq 0$ for all p or $A_p \leqq 0$ for all p; but we have

$$\arg \prod_{k \neq p} (\rho e^{\phi i} - \tau_k) = \sum_{k \neq p} \phi_k = \Psi_p,$$

and the theorem is established.

THEOREM 70. *If $E_s(\rho)$ is the set of points of the z-plane at which $F_{\rho\sin}$ is served by $+T_n(x)$ or $-T_n(x)$ then a necessary and sufficient condition that a point z belongs to E_s is that for this point either the $(\Psi_p)_0^n$ are in $[0, \pi]$ (and then $+T_n(x)$ serves), or they are in $[-\pi, 0]$ (and then $-T_n(x)$ serves).*

In fact, in this case we have

$$\delta_p = (-1)^{n-p} \frac{\frac{1}{2i}\left[\prod_{k\neq p} (\rho e^{\phi i} - \tau_k) - \prod_{k \neq p} (\rho e^{-\phi i} - \tau_k)\right]}{\prod_{k \neq p} |\tau_p - \tau_k|} \qquad (p = 0, 1, \cdots, n).$$

Here the numerator $B_p = \operatorname{Im} \prod_{k \neq p} (\rho e^{\phi i} - \tau_k)$. We must have either $B_p \geqq 0$ for all p, or $B_p \leqq 0$ for all p, i.e. the Ψ_p must lie either all in the upper half-plane or all in the lower half-plane.

We shall use the term Cebyšev arcs for the arcs of the circles of radius $\rho =$ const. at each point of which $F_{\rho\cos}$ (or $F_{\rho\sin}$) is served by one of the polynomials $\pm T_n(x)$.

COROLLARY. *The arguments of the endpoints of the Čebyšev arcs* ($\rho =$ const.) *are obtained when* $\Psi_0(\phi)$ *or* $\Psi_n(\phi)$ *is tangent to the imaginary axis. Denote these arguments by* $a < b$; *since* $\Psi_0(\phi) > \Psi_n(\phi)$ *for every* ϕ *we have* $\delta_0 = 0$ *for* $\phi = a$ *and we infer that* $\phi = a$ *is one of the endpoints of a Čebyšev arc at which the weight at* $\tau_0 = 0$ *is lost; for* $\phi = b$ *the node* $\tau_n = 1$ *is lost* (*i.e.* $\delta_n = 0$); *service at these endpoints is by a polynomial of passport* $[n, n, 0]$.

The impossibility of serving $F_{\rho\cos}$, $F_{\rho\sin}$ by polynomials of passport $[n, n, 1]$, also containing n nodes, follows from the theorem on continuous deformation, which requires that at points where the passport changes the functional loses still another node, but then $s < n$, which is inadmissible.

There is a similar result for $F_{\rho\sin}$, where also one of the boundary nodes of the Čebyšev polynomial is lost at the endpoints of a Čebyšev arc.

Hence in the whole z-plane $F_{\rho\cos}$ and $F_{\rho\sin}$ are served either by Čebyšev polynomials or by polynomials of passport $[n, n, 0]$. This establishes the stability of both functionals.

As in Chapter IV of Part 1 we denote the Zolotarev polynomials by $Q_n(x, \vartheta) = \vartheta x^n + y_{n-1}(\vartheta) x^{n-1} + \cdots$. We shall find the polynomials serving the segments $F_{\rho\cos}$ and $F_{\rho\sin}$ at the corresponding Zolotarev points. We denote the sets of these points in the z-plane by $\Xi_c^{(\rho)}$ and $\Xi_s^{(\rho)}$. Then if $z \in \Xi_c^{(\rho)}$ the value of ϑ is determined by the condition

$$\vartheta\rho^n \cos n\phi + y_{n-1}(\vartheta)\rho^{n-1}\cos(n-1)\phi + \cdots + y_0(\vartheta) = \max(\vartheta),$$

i.e.

$$\rho^n \cos n\phi + y'_{n-1}(\vartheta)\rho^{n-1}\cos(n-1)\phi + \cdots + y'_0(\vartheta) = 0.$$

If $z \in \Xi_s^{(\rho)}$ then ϑ is determined by the condition

$$\vartheta\rho^n \sin n\phi + y_{n-1}(\vartheta)\rho^{n-1}\sin(n-1)\phi + \cdots + y_1(\vartheta)\sin\phi = \max(\vartheta),$$

i.e.

$$\rho^n \sin n\phi + y'_{n-1}(\vartheta)\rho^{n-1}\sin(n-1)\phi + \cdots + y'_1(\vartheta)\sin\phi = 0.$$

In the special case $\rho = 1$ we obtain estimates for real trigonometric polynomials of the forms

$$\sum a_k \cos k\phi = C_n(\phi) \quad \text{and} \quad \sum b_k \sin k\phi = S_n(\phi);$$

in fact

$$|C_n(\phi)| \leqq |F_{\cos}(T(x))| \cdot \max|\sum a_k x^k| \text{ for } \phi \in E_c^{(1)};$$
$$|S_n(\phi)| \leqq |F_{\sin}(T(x))| \cdot \max|\sum b_k x^k| \text{ for } \phi \in E_s^{(1)};$$
$$|C_n(\phi)| \leqq |F_{\cos}[Q_n(x,\vartheta)]| \cdot \max|\sum a_k x^k| \text{ for } \phi \in \Xi_c^{(1)};$$
$$|S_n(\phi)| \leqq |F_{\sin}[Q_n(x,\vartheta)]| \cdot \max|\sum b_k x^k| \text{ for } \phi \in \Xi_s^{(1)}.$$

The estimates are sharp, since the upper bounds are attained for the extremal polynomials.

The arcs of a circle of radius ρ that lie between two successive Čebyšev arcs are called Zolotarev arcs. We denote the arguments of their endpoints by α_k and β_k ($\alpha_k < \beta_k$). The number of these arcs and the arguments of their endpoints on the semicircle $(0, \pi)$ are determined for any $\rho \geqq 1$ and for $F_{\rho\cos}$ by

$$\Psi_0 = \frac{\pi}{2}, \frac{3\pi}{2}, \cdots, (2n-1)\frac{\pi}{2} \text{ for } \alpha_k$$

and

$$\Psi_n = \frac{\pi}{2}, \frac{3\pi}{2}, \cdots, (2n-1)\frac{\pi}{2} \text{ for } \beta_k.$$

Consequently the semicircle always contains n Zolotarev arcs.

§2. Asymptotic behavior of the Zolotarev regions in the z-plane [21]

We note some elementary formulas for arbitrary $z = \rho e^{\phi i}$ in the upper half-plane. Put $\bar{\phi}_k = \phi_k - \phi_0 = \Psi_0 - \Psi_k$. Then

$$\bar{\phi}_k = \frac{\pi}{2} - \frac{\phi_0}{2} - \operatorname{arc\,tg}\left[\frac{\rho - r_k}{\rho + r_k}\operatorname{ctg}\frac{\phi_0}{2}\right], \tag{118}$$

and in addition

$$\psi_0 = n\phi_0 + \sum_{k=1}^{n} \bar{\phi}_k, \tag{119}$$

$$\psi_n = n\phi_0 + \sum_{k=1}^{n} \bar{\phi}_k - \frac{\pi}{2} + \frac{\phi_0}{2} + \operatorname{arc\,tg}\left[\frac{\rho - 1}{\rho + 1}\operatorname{ctg}\frac{\phi_0}{2}\right]. \tag{120}$$

It is clear that

$$\max_{(\phi_0)} \bar{\phi}_k = \arcsin\frac{r_k}{\rho} \quad \left(k = \text{const.}, \ \phi_k = \frac{\pi}{2}, \ \rho > 1\right).$$

The largest term in $\sum_1^n \bar{\phi}_k$ is $\bar{\phi} \leqq \arcsin(1/\rho)$; hence (for $n = \text{const.}$) we have

(121) $$\sum_1^n \bar{\phi}_k < n \arcsin \frac{1}{\rho}, \quad \lim_{\rho\to\infty} \sum_1^n \bar{\phi}_k = 0.$$

THEOREM 71. *On the semicircle $0 < \phi_0 < \pi$, as ρ increases the arguments of the endpoints of the kth Zolotarev arc ($\alpha_k < \beta_k$) approach $(2k-1)\pi/2n$.*

In fact, according to the Corollary of Theorem 70 we obtain from (119) with $\phi_0 = \alpha_k$

$$n\alpha_k + \sum_{p=1}^n \bar{\phi}_p(\alpha_k) = \frac{2k-1}{2}\pi,$$

and from (120) with $\phi_0 = \beta_k$

$$n\beta_k + \sum_{p=1}^n \bar{\phi}_p(\beta_k) - \frac{\pi}{2} + \frac{\beta_k}{2} + \operatorname{arc\,tg}\left[\frac{\rho-1}{\rho+1}\operatorname{ctg}\frac{\beta_k}{2}\right] = \frac{(2k-1)\pi}{2}.$$

By (121) we obtain

$$\lim_{\rho\to\infty} \alpha_k = \lim_{\rho\to\infty} \beta_k = \frac{(2k-1)\pi}{2n}.$$

THEOREM 72. *In the z-plane the branches $\rho e^{\alpha_k(\rho)i}$ of the boundary of the Zolotarev regions have a system of asymptotes issuing from the point $(n+1)/2n$ on the x-axis; and for the branches $\rho e^{\beta_k(\rho)i}$ the asymptotes are lines issuing from the point $(n-1)/2n$ (the slopes of the asymptotes are determined in Theorem 71).*

In fact, suppose that $l(\alpha_k)$ (or $l(\beta_k)$) is the length of the arc of radius ρ connecting the points $\rho e^{(2k-1)\pi i/2n}$ and $\rho e^{\alpha_k i}$ (or $\rho e^{\beta_k i}$); then we have (by the formulas of Theorem 71):

$$l(\alpha_k) = \rho\left(\frac{2k-1}{2n}\pi - \alpha_k\right) = \frac{\rho}{n}\sum_{p=1}^n \bar{\phi}_p(\alpha_k);$$

$$l(\beta_k) = \rho\left(\frac{2k-1}{2n}\pi - \beta_k\right) = \frac{\rho}{n}\sum_{p=1}^{n-1} \bar{\phi}_p(\beta_k).$$

The oblique abscissa of the point $\rho e^{\alpha_k i}$ is

$$x_\alpha = \frac{\rho \sin\left(\dfrac{2k-1}{2n}\pi - \alpha_k\right)}{\sin\dfrac{2k-1}{2n}\pi} = \rho\frac{\sin\left[\dfrac{1}{n}\sum_1^n \bar{\phi}_p(\alpha_k)\right]}{\sin\dfrac{2k-2}{2n}\pi}.$$

Here we may proceed to the limit

$$\lim_{\rho\to\infty}\left[\rho\sin\frac{1}{n}\sum_1^n\overline{\phi}_p(\phi)\right]$$

with $\phi = \text{const.}$, and then put $\phi = (2k-1)\pi/2n$. Since

$$\sum_{p=1}^n \overline{\phi}_p(\phi)\to 0,$$

we have

$$\begin{aligned}\lim\frac{\rho}{n}\sum_1^n\overline{\phi}_p(\phi) &= \frac{1}{n}\lim\rho\sum_1^n\left(\frac{\pi}{2}-\frac{\phi}{2}-\operatorname{arc\,tg}\frac{\rho-\tau_k}{\rho+\tau_k}\operatorname{ctg}\frac{\phi}{2}\right)\\ &= \frac{1}{n}\sum_{k=1}^n\tau_k\frac{2\operatorname{ctg}(\phi/2)}{1+\operatorname{ctg}^2(\phi/2)} = \frac{1}{n}\sin\phi\sum_1^n\tau_k.\end{aligned}$$

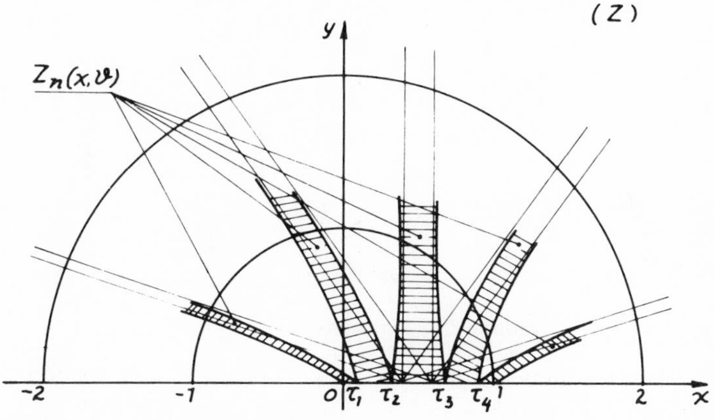

FIGURE 7. Topography of $F_{\cos}$ for $n = 5$

Hence $\lim x_\alpha = (n+1)/2n$. Similarly we obtain $\lim x_\beta = (n-1)/2n$, and the theorem is established (Figure 7).

COROLLARY. *The sum of the Zolotarev arcs on the semicircle of radius ρ approaches $1/(n\sin(\pi/2n))$ as $\rho\to\infty$.*

In fact, the length of the kth Zolotarev arc is

$$l(\alpha_k)-l(\beta_k) = \frac{\rho}{n}\left[\sum_{p=1}^n\overline{\phi}_p(\alpha_k)-\sum_{p=1}^{n-1}\overline{\phi}_p(\beta_k)\right]\underset{\rho}{\to}\frac{1}{n}\sin\frac{(2k-1)\pi}{2n}.$$

For the sum we have in the limit

$$\frac{1}{n}\sum_{k=1}^n\sin\frac{2k-1}{2n}\pi = \frac{1}{n\sin(\pi/2n)}.$$

Hence for large n this sum is close to $2/\pi$.

For $F_{\sin}$ the same process leads to similar results. As $\rho \to \infty$ the arguments of the endpoints of the Zolotarev arc (α_k, β_k) tend to π/n $(k = 1, 2, \cdots, n-1)$; the nth Zolotarev arc reduces, for arbitrary ρ, to the point $\alpha_n = \beta_n = \pi$; as $\rho \to \infty$ the branches $\rho e^{\alpha_k(\rho)i}$ and $\rho e^{\beta_k(\rho)i}$ have asymptotes issuing from the points $(n+1)/2n$ and $(n-1)/2n$ of the x-axis; the sum of the lengths of the Zolotarev arcs approaches $n^{-1}\cot(\pi/2n)$ as $\rho \to \infty$ (Figure 8).

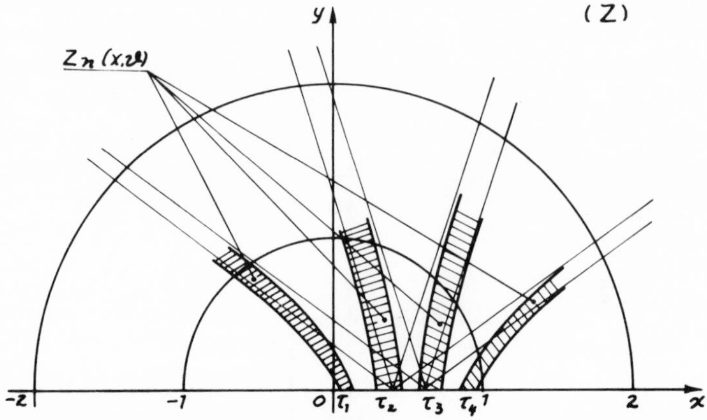

FIGURE 8. Topography of $F_{\sin}$ for $n = 5$

REMARK. It follows from the properties we have established that the Zolotarev arcs for $F_{\cos}$ and $F_{\sin}$ for the same $\rho \geqq 1$ do not overlap.

To discuss the boundaries of the Zolotarev regions for $\rho < 1$, we note, considering only the case of $F_{\cos}$, that in general the arguments of the endpoints $\alpha(\rho)$ and $\beta(\rho)$ are roots of the equations

$$F_{\cos}\left[\frac{R_{n+1}(x)}{x}\right] = 0; \quad F_{\cos}\left[\frac{R_{n+1}(x)}{x-1}\right] = 0,$$

where $R_{n+1}(x) = \prod_0^n (x - \tau_k)$. Putting

$$\frac{R_{n+1}}{x} = \sum_0^n S_{n-k}^{(1)} x^k, \quad \frac{R_{n+1}(x)}{x-1} = \sum_0^n S_{n-k}^{(2)} x^k,$$

we have (for any $\rho > 0$)

$$\rho^n \cos n\alpha + S_1^{(1)} \rho^{n-1} \cos(n-1)\alpha + \cdots + S_n^{(1)} = 0,$$

$$\rho^n \cos n\beta + S_1^{(2)} \rho^{n-1} \cos(n-1)\beta + \cdots + S_{n-1}^{(2)} \rho \cos\beta = 0.$$

Putting $\alpha = 0$ or $\beta = 0$, we have

$$\prod_1^n (\rho - \tau_k) = 0 \quad \text{or} \quad \prod_0^{n-1} (\rho - \tau_k) = 0.$$

Therefore the curves $\rho e^{\alpha(\rho)i}$ and $\rho e^{\beta(\rho)i}$ intersect the x-axis at the points $(\tau_p)_1^n$ and $(\tau_p)_0^{n-1}$, and the Čebyšev regions have "sources" on $[0,1]$ at the points $(\tau_k)_0^n$, whereas the Zolotarev regions contain the whole interval $[0,1]$ except for (τ_k). The sources for the α_k and β_k branches are τ_{n-k+1} and τ_{n-k}, respectively.

We note an immediate application of our theorems to classical extremal problems. Let $E_{\cos}$ and $E_{\sin}$ be the Čebyšev sets in the z-plane for $F_{\cos}$ and $F_{\sin}$, and let $\Xi_{\cos}$ and $\Xi_{\sin}$ be the Zolotarev sets.

1. Among all algebraic polynomials $\{P_n(x)\}$ with real coefficients, reduced on $[0,1]$, find the one that has maximum modulus at the point z_0. If $z_0 \in (E_{\cos}, E_{\sin})$, this polynomial is clearly $\pm T_n(x)$, and $|T_n(z_0)| = N$ is the norm of the functional $(z_0^k)_{k=0}^n$.

Problem 1 can be reformulated as follows:

1A. Among all $\{P_n(x)\}$ such that $|P_n(z_0)| = A$, find the one that deviates least from zero on $[0,1]$. If the solution of Problem 1A is $Y_n(x)$ with deviation L, and $P_n(x)$ is the solution of Problem 1, then $Y_n(x)/L = P_n(x)$; and $A = LN$.

2. Among all $\{P_n(x)\}$, reduced on $[0,1]$, find the one that yields at z_0: a) $\max|\operatorname{Re} P_n(z)|$, or b) $\max|\operatorname{Im} P_n(z)|$. If $z_0 \in (E_{\cos}, E_{\sin})$, the required polynomial is $\pm T_n(x)$ for either a) or b). If $z_0 \in \Xi_{\cos}$, the polynomial for a) belongs to the family $Z_n(x,\vartheta)$, where ϑ is the leading coefficient; for $z_0 = \rho e^{\phi i}$ the value of ϑ is determined from $F_{\cos}[Z_n(x,\vartheta)] = \max(\vartheta)$, i.e. $\partial F_{\cos}(Z_n)/\partial\vartheta = 0$ [7], or, what is the same thing, $F_{\cos}[R_n(Z)] = 0$, where $R_n(x) = \prod_1^n (x - \sigma_k)$; $\sigma_k(\vartheta)$ are the nodes of the Zolotarev polynomial; in this case the extremal polynomial for b) is $\pm T_n(x)$.

There is a similar solution for $Z_0 \in \Xi_{\sin}$.

Problem 2 can also be reformulated: among all $\{P_n(x)\}$ for which $|\operatorname{Re} P_n(z_0)| = A$ or $|\operatorname{Im} P_n(z_0)| = B$, find the one deviating least from zero on $[0,1]$.

In conclusion we formulate some very simple but still unsolved problems related to the problems discussed in this monograph.

1. Considering the extremal polynomials $\{Q_n(x)\}$ of class II to be known, construct the corresponding reduced trigonometric polynomials and apply them to the solution of Čebyšev problems.

We have solved this problem for cosine and for sine polynomials. The general case has not been investigated. (See [**25**].)

2. Investigate the functional

$$F_\xi = 1, \frac{\xi}{2}, \frac{\xi^2}{3}, \frac{\xi^3}{4}, \cdots, \frac{\xi^n}{n+1} \quad \text{for} \quad -\infty < \xi < +\infty.$$

It is natural to call F_ξ the integral functional:

$$F_\xi(P_n) = \frac{1}{\xi}\int_0^\xi P_n(x)\,dx.$$

We are to find its extremal polynomial $Q_n(x, \xi)$; the investigation is needed only for very small intervals of ξ, since outside them the behavior of the functional can be obtained very simply. (See [**34**].)

3. In determining the best approximation to a given analytic function $f(x)$ by polynomials of given degree (Chapter II of Part 2) it was necessary to say "for sufficiently large N." It would be desirable to find upper and lower bounds for N, depending on the nature of $f(x)$.

4. Complete the problems of Chapter IV by finding the exact majorant for a reduced $Q_n(x)$ with real coefficients throughout the z-plane, i.e. for $Q_n(z)$.

Here it is a question of investigating the cases when z is not in both $E_{\cos}$ and $E_{\sin}$. Is there then an extremal polynomial differing both from $\pm T_n(x)$ and from polynomials of passport $[n, n, 0]$?

5. In the problem of Chapter II of Part 2 (§1), investigate the case $m < n/2$ and construct the unique solution of the problem, an extremal polynomial of class I.

BIBLIOGRAPHY

1. I. P. Natanson, *Theory of functions of a real variable,* GITTL, Moscow, 1957; English transl., Ungar, New York, 1961. MR **26** # 6309.

2. ———, *Constructive function theory,* GITTL, Moscow, 1949; English transl., Ungar, New York, 1964. MR **11**, 591; MR **33** # 4529a.

3. F. Hausdorff, *Summationsmethoden und Momentfolgen,* Math. Z. **9** (1921), 74-109, 280-299.

4. E. V. Voronovskaja, *On the transformation of a functional series by differences of its terms,* Dokl. Akad. Nauk SSSR (A) **1930**, 693-700. (Russian)

5. ———, *A minimal problem in the theory of absolutely monotonic sequences,* Trudy Leningrad. Indust. Inst. Razdel Fiz.-Mat. Nauk **4** (1937), no. 2, 22-33. (Russian)

6. ———, *The norming of finite functionals.* I, Trudy Leningrad. Indust. Inst. Razdel Fiz.-Mat. Nauk **5** (1938), no. 1, 14-20. (Russian)

7. ———, *Extremal polynomials of finite functionals,* Thesis Abstract, Leningrad. Gos. Univ., Leningrad, 1955. (Russian)

8. ———, *The norming of finite functionals.* II, Trudy Leningrad. Politehn. Inst. **3** (1941), 23-33. (Russian)

9. L. V. Kantorovič and G. P. Akilov, *Functional analysis in normed spaces,* Fizmatgiz, Moscow, 1959; English transl., Internat. Series of Monographs in Pure and Appl. Math., vol. 46, Macmillan, New York, 1964. MR **22** # 9837; MR **35** # 4699.

10. P. L. Čebyšev, Complete collected works. Vols. II, III, Izdat. Akad. Nauk SSSR, Moscow, 1947, 1948. (Russian) MR **11**, 150.

11. E. I. Zolotarev, *Application of elliptic functions to problems on functions deviating least or most from zero,* Zapiski S.-Peterburg. Akad. Nauk **30** (1877), no. 5; reprinted in Collected works. Vol. II, Izdat. Akad. Nauk SSSR, Leningrad, 1932, pp. 1-59. (Russian)

12. V. A. Markov, *On functions deviating least from zero in a given interval,* Izdat. Akad. Nauk, St. Petersburg, 1892. (Russian)

13. A. P. Pšeborskiĭ, *On certain polynomials deviating least from zero in a given interval,* Soobšč. Har'kov. Mat. Obšč. (2) **14** (1913), 65-80. (Russian)

14. E. V. Voronovskaja, *On the closest uniform approximation of polynomials,* Dokl. Akad. Nauk SSSR **114** (1957), 927-929. (Russian) MR **20** # 1151.

15. V. L. Gončarov, *Theory of interpolation and approximation of functions,* 2nd ed., GITTL, Moscow, 1954. (Russian) MR **16**, 803.

16. E. V. Voronovskaja, *On Čebyšev's approximation of analytic functions by algebraic polynomials,* Dokl. Akad. Nauk SSSR **121** (1958), 206-209. (Russian) MR **20** # 6622.

17. A. A. Markov, *On a problem of D. I. Mendeleev,* Zap. Akad. Nauk St. Petersburg **62** (1889), 1-24. (Russian)

18. S. N. Bernšteĭn, Collected works. Vol. I, Izdat. Akad. Nauk SSSR, Moscow, 1952. (Russian) MR **14**, 2.

19. V. A. Gusev, *Derivative functionals of algebraic polynomials and V. A. Markov's theorem,* Izv. Akad. Nauk SSSR Ser. Mat. **25** (1961); English transl., Appendix: Transl. Math. Monographs, vol. 28, Amer. Math. Soc., Providence, R. I., 1970. MR **14** # A797.

20. E. V. Voronovskaja, *The structure of trigonometric functionals and their properties,* Uspehi Mat. Nauk **12** (1957), no. 5 (77), 254-257. (Russian)

21. E. V. Voronovskaja and M. Ja. Zinger, *An estimate for polynomials in the complex plane,* Dokl. Akad. Nauk SSSR **143** (1962), 1022-1025 = Soviet Math. Dokl. **3** (1962), 516-519. MR **25** # 3149.

22. E. V. Voronovskaja, *The functional of the first derivative and improvement of a theorem of A. A. Markov,* Izv. Akad. Nauk SSSR Ser. Mat. **23** (1959), 951-962. MR **22** # 1640.
23. R. G. Selfridge, *Approximations with least maximum error,* Pacific J. Math. **3** (1953), 247-255. MR **14**, 1067.
24. S. N. Bernšteĭn, *Extremal properties of polynomials and best approximation of continuous functions of a real variable,* Glaz. Redak. Obšč. Lit., Leningrad, 1937. (Russian)
25. E. V. Voronovskaja, *Extremal trigonometric polynomials and their applications,* Dokl. Akad. Nauk SSSR **129** (1959), 12-15. (Russian) MR **22** # 1782.

The references from here on are to relevant papers appearing after the publication of the book and selected by the translator.

26. E. V. Voronovskaja and M. Ja. Zinger, *An estimate for polynomials in the complex plane,* Dokl. Akad. Nauk SSSR **143** (1962), 1022-1025 = Soviet Math. Dokl. **3** (1962), 516-519. MR **25** # 3149.
27. E. V. Voronovskaja, *Odd polynomials of least deviation,* Dokl. Akad. Nauk SSSR **159** (1964), 715-718 = Soviet Math. Dokl. **5** (1964), 1562-1565. MR **31** # 1507.
28. ———, *Some criteria for the stability of functionals,* Dokl. Akad. Nauk SSSR **161** (1965), 270-273 = Soviet Math. Dokl. **6** (1965), 392-395. MR **31** # 2544.
29. M. Ja. Zinger, *Functionals of derivatives in the complex plane,* Dokl. Akad. Nauk SSSR **166** (1966), 775-778 = Soviet Math. Dokl. **7** (1966), 158-161. MR **33** # 5855.
30. ———, *Generalization of a problem of Schaeffer-Duffin to finite functionals,* Dokl. Akad. Nauk SSSR **172** (1967), 14-17 = Soviet Math. Dokl. **8** (1967), 7-10. MR **35** # 635.
31. E. V. Voronovskaja, *On Zolotarev stability of functionals,* Dokl. Akad. Nauk SSSR **173** (1967), 15-17 = Soviet Math. Dokl. **8** (1967), 304-306. MR **35** # 4646.
32. ———, *Amorphous functionals, their structure and applications,* Dokl. Akad. Nauk SSSR **166** (1966), 1270-1273 = Soviet Math. Dokl. **7** (1966), 258-262. MR **35** # 5841.
33. M. Ja. Zinger, *An estimate of the derivatives of an algebraic polynomial in the complex plane,* Sibirsk. Mat. Ž. **8** (1967), 952-957. (Russian) MR **36** # 332.
34. V. A. Gusev, *An integral functional and its norm,* Dokl. Akad. Nauk SSSR **168** (1966), 13-16 = Soviet Math. Dokl. **7** (1966), 575-578. MR **36** # 6839.
35. E. V. Voronovskaja, *The construction of extremal trigonometric polynomials of mixed type,* Dokl. Akad. Nauk SSSR **179** (1968), 768-771 = Soviet Math. Dokl. **9** (1968), 438-441. MR **37** # 3267.
36. ———, *Structural properties of certain determining functionals,* Dokl. Akad. Nauk SSSR **180** (1968), 1271-1274 = Soviet Math. Dokl. **9** (1968), 737-741. MR **38** # 284.
37. M. Ja. Zinger, *Estimates of the derivatives of an algebraic polynomial,* Sibirsk. Mat. Ž. **10** (1969), 318-328 = Siberian Math. J. **10** (1969), 225-232. MR **40** # 605.
38. I. Ju. Ryžakov, *The trigonometrical analogue of a problem of E. I. Zolotarev,* Izv. Vysš. Učebn. Zaved. Matematika 1969, no. 8 (87), 75-88. (Russian)

APPENDIX

DERIVATIVE FUNCTIONALS OF AN ALGEBRAIC POLYNOMIAL AND V. A. MARKOV'S THEOREM[1]

V. A. GUSEV

The functional method is applied to investigate the exact upper bounds of the derivatives of order k of an algebraic polynomial at each point of the interval $[0,1]$. A new proof of V. A. Markov's theorem is given and some new results are obtained.

V. A. Markov's theorem [1][2] gives a bound for the derivative of order $k \leqq n$ of an algebraic polynomial $P_n(x)$: if

$$\max_{[-1,1]} |P_n(x)| = M, \tag{1}$$

then

$$\max_{[-1,1]} |P_n^{(k)}(x)| \leqq \frac{n^2(n^2-1^2)\cdots(n^2-\overline{k-1}^2)}{1\cdot 3\cdots(2k-1)}\cdot M \quad (k=1,2,\cdots,n). \tag{2}$$

New and simpler proofs of this theorem were given by S. N. Bernšteĭn [2] and Duffin and Schaeffer [3].

The functional method developed by E. V. Voronovskaja [4]—[6] makes it possible to study the problem considerably more completely than in [2] and [3], and in a considerably shorter way than in [1].

The case $k=1$ was studied by E. V. Voronovskaja [7]. In the present paper the results obtained in [7] for the first derivative are extended to derivatives of higher order; the terminology of [7] is retained here.

We consider the interval $[0,1]$ (this is the most convenient for the functional method). Then (2) takes the form

$$\max_{[0,1]} |P_n^{(k)}(x)| \leqq \frac{2^k n^2(n^2-1^2)\cdots(n^2-\overline{k-1}^2)}{1\cdot 3\cdots(2k-1)}\cdot M_{[0,1]} = M_{[0,1]}T_n^{(k)}(1), \tag{2'}$$

where

$$M_{[0,1]} = \max_{[0,1]} |P_n(x)|, \quad T_n(x) = \cos n \arccos(2x-1).$$

---

[1] Translation of Izv. Akad. Nauk SSSR Ser. Mat. **25** (1961), 367-384.

[2] Numbers in brackets refer to the Bibliography at the end of this Appendix.

We shall treat the kth derivative of $P_n(x)$ at a point ξ as the linear functional $F_\xi^{(k)}$, defined by the finite sequence (or segment)

$$(\mu_i^{(k)})_{i=0}^n = 0_0, 0_1, \cdots, 0_{k-1}, k!, (k+1)!\xi, \cdots, \frac{n!}{(n-k)!}\xi^{n-k} \tag{3}$$

$$(k = 1, 2, \cdots, n; \ -\infty < \xi < \infty),$$

on the set of polynomials $\{P_n(x)\}$ of degree at most n, so that

$$F_\xi^{(k)}(x^i) = 0 \qquad \text{for } i = 0, 1, \cdots, k-1,$$

$$F_\xi^{(k)}(x^i) = \frac{i!}{(i-k)!}\xi^{i-k} \text{ for } i = k, k+1, \cdots, n$$

and

$$F_\xi^{(k)}[P_n(x)] = P_n^{(k)}(\xi).$$

Since the space $\{P_n(x)\}$ is finite-dimensional, for each ξ there is a polynomial $Q_n(x, \xi)$, called extremal or serving, such that when

$$\max_{[0,1]} |Q_n(x, \xi)| = 1$$

(reduced polynomial) we have the equation

$$N_k(\xi) = [Q_n^{(k)}(x, \xi)]_{x=\xi},$$

where $N_k(\xi)$ is the norm of the functional (3). We have

$$\max_{[0,1]} |P_n^{(k)}(\xi)| \leqq M_{[0,1]} \cdot N_k(\xi) = M_{[0,1]} \cdot [Q_n^{(k)}(x, \xi)]_{x=\xi} \quad (k = 1, 2, \cdots, n).$$

This is a sharpening of V. A. Markov's inequality (2′). Thus the problem is to determine the extremal polynomial and the norm of the functional (3).

REMARK 1. Putting $\mu_i^{(k)} = \mu_{i,0}^{(k)}$, we form the table of differences $(\mu_{i,p}^{(k)})$ of the segment (3), where

$$\mu_{l,p}^{(k)} = \mu_{l,p-1}^{(k)} - \mu_{l+1,p-1}^{(k)};$$

then we obtain

$$(\mu_{0j}^{(k)}) = 0_0, 0_1, \cdots, 0_{k-1}, (-1)^k k!, (-1)^k (k+1)!(1-\xi), \cdots$$

$$\cdots, \frac{(-1)^k n!}{(n-k)!}(1-\xi)^{n-k}.$$

This functional has the same norm (cf. [4]) as (3). Consequently

$$N_k(\xi) = N_k(1-\xi) \quad (k = 1, 2, \cdots, n; \ -\infty < \xi < \infty). \tag{4}$$

REMARK 2. For $\xi < 0$ the segment (3) has alternating signs, and consequently

$$Q_n(x, \xi) = + T_n(x),$$

if $k \equiv n \pmod 2$, and

$$Q_n(x, \xi) = - T_n(x),$$

if $k + 1 \equiv n \pmod 2$. In view of (4), for each ξ outside $[0, 1]$ we have

$$Q_n(x, \xi) = \pm T_n(x), \quad N_k(\xi) = | T_n^{(k)}(\xi) | \quad (k = 1, 2, \cdots, n).$$

We recall a necessary and sufficient condition that a given reduced polynomial $L_n(x)$ $(\max_{[0,1]} | L_n(x) | = 1)$ with distribution $(\overset{\pm}{\sigma}_i)_1^s$ $(0 \leqq \sigma_1 < \sigma_2 < \cdots < \sigma_s \leqq 1$; $L_n(\overset{+}{\sigma}_i) = + 1, L_n(\bar{\sigma}_i) = - 1)$ is extremal for a given segment $(\mu_i)_0^n$. The test for extremal character is as follows (cf. [7]): the system of $n + 1$ equations in the s unknowns δ_i

$$\text{(V)} \qquad \sum_{i=1}^{s} \delta_i \sigma_i^l = \mu_l \quad (l = 0, 1, \cdots, n)$$

has the properties 1) it is consistent, and 2) it has a solution such that either $\operatorname{sgn} \delta_i = L_n(\sigma_i)$ or $\delta_i = 0$ (but not all are zero).

THEOREM 1. *On the interval* $[0, 1]$ *there are* $n - k + 1$ *Čebyšev intervals* $[\alpha_i^{(k)}, \beta_i^{(k)}]$ $(k = 1, 2, \cdots, n;\ i = 1, 2, \cdots, n - k + 1)$ *at whose points the functional* (3) *is served by one of the polynomials* $\pm T_n(x)$. *The endpoints of the intervals are the roots* $\alpha_2^{(k)}, \alpha_3^{(k)}, \cdots, \alpha_{n-k+1}^{(k)}$ *of the equation*

$$\frac{d^k}{dx^k} \left(\frac{R_{n+1}(x)}{x} \right) = 0,$$

and $\alpha_1^{(k)} = 0$, *and the roots* $\beta_1^{(k)}, \beta_2^{(k)}, \cdots, \beta_{n-k}^{(k)}$ *of the equation*

$$\frac{d^k}{dx^k} \left(\frac{R_{n+1}(x)}{x - 1} \right) = 0,$$

and $\beta_{n-k+1}^{(k)} = 1$; *here* $R_{n+1}(x) = \prod_{i=0}^{n} (x - \tau_i)$ *is the resolvent of* $T_n(x)$, *and* $(\tau_i)_0^n$ *are its nodes, i.e. its points of maximum deviation from zero.*

PROOF. Let us apply the test for extremal character to $L_n(x) = \pm T_n(x)$. In this case $s = n + 1$ and condition 1) drops out, while condition 2) is that $\delta_0, \delta_1, \cdots, \delta_n$ have alternating signs. The system (V) for the segment (3) takes the form

$$\sum_{i=0}^{n} \delta_i \tau_i^l = 0 \qquad (l = 0, 1, \cdots, k - 1),$$

$$\sum_{i=0}^{n} \delta_i \tau_i^l = \frac{l!}{(l - k)!} \xi^{l-k} \qquad (l = k, k + 1, \cdots, n),$$

and its solution is given by

$$\delta_j = \frac{(-1)^{n-j}}{\prod_{i\neq j}|\tau_j - \tau_i|} \cdot \left[\frac{d^k}{dx^k}\left(\frac{R_{n+1}(x)}{x-\tau_j}\right)\right]_{x=\xi} \qquad (j = 0, 1, \cdots, n). \tag{5}$$

We shall need the following three lemmas of V. A. Markov [1].

LEMMA 1. *If the equation* $G_s(x) = x^s + a_1 x^{s-1} + \cdots + a_{s-1}x + a_s = 0$ *has no complex roots, then*

$$[G_s^{(k)}(x)]^2 - G_s^{(k-1)}(x) \cdot G_s^{(k+1)}(x) > 0$$

for all $k \leq s$ *and all real* x *except for roots of* $G_s(x) = 0$ *of multiplicity greater than* k.

LEMMA 2. *Let* $G(x) = \prod_{i=1}^s (x - x_i)$ $(x_k \neq x_j)$, *let* z *be a root of* $G^{(k)}(x) = 0$ *and let*

$$G_l(x) = \frac{G(x)}{x - x_l} \qquad (l = 1, 2, \cdots, s);$$

then the numbers $G_1^{(k)}(z), G_2^{(k)}(z), \cdots, G_s^{(k)}(z), G^{(k+1)}(z)$ *all have the same sign.*

LEMMA 3. *Let*

$$G(x) = A \cdot \prod_{i=1}^s (x - a_i), \quad H(x) = B \cdot \prod_{i=1}^s (x - b_i) \quad (A > 0,\ B > 0)$$

and $b_1 < a_1 < b_2 < a_2 < \cdots < b_s < a_s$. *If* $G^{(k)}(z) = 0$ *then*

$$\frac{H^{(k)}(z)}{G^{(k+1)}(z)} > 0.$$

COROLLARY. *It follows immediately from Lemma 3 that the roots of* $G^{(k)}(x) = 0$ *and of* $H^{(k)}(x) = 0$ *are interlaced. This is also true when the degrees of* $G(x)$ *and* $H(x)$ *differ by unity (the roots of* $G(x)$ *and* $H(x)$ *separate each other).*

Let $\Phi_j(x)$ be the polynomial

$$\frac{R_{n+1}(x)}{x - \tau_j} \qquad (j = 0, 1, \cdots, n)$$

and let $\xi_{j,k}^{(1)}, \xi_{j,k}^{(2)}, \cdots, \xi_{j,k}^{(n-k)}$ be the roots of $\Phi_j^{(k)}(x)$; then $\xi_{0,k}^{(i)} = \alpha_{i+1}^{(k)}$ and $\xi_{n,k}^{(i)} = \beta_i^{(k)}$; let $\theta_1^{(k)}, \theta_2^{(k)}, \cdots, \theta_{n-k+1}^{(k)}$ be the roots of $R_{n+1}^{(k)}(x)$. It follows from the results of Voronovskaja for $k = 1$ (cf. [7]) that the following inequalities hold for the roots of $\Phi_0'(x), \Phi_1'(x), \cdots, \Phi_n'(x)$ and $R_{n+1}'(x)$:

$$0 < \theta_1^{(1)} < \beta_1^{(1)} < \xi_{n-1,1}^{(1)} < \xi_{n-2,1}^{(1)} < \cdots < \xi_{1,1}^{(1)} < \alpha_2^{(1)} < \theta_2^{(1)} < \beta_2^{(1)} < \xi_{n-1,1}^{(2)} < \cdots$$
$$\cdots < \xi_{1,1}^{(2)} < \alpha_3^{(1)} < \theta_3^{(1)} < \beta_3^{(1)} < \cdots < \alpha_{n-1}^{(1)} < \theta_{n-1}^{(1)} < \beta_{n-1}^{(1)} < \xi_{n-1,1}^{(n-1)} < \cdots$$
$$\cdots < \xi_{1,1}^{(n-1)} < \alpha_n^{(1)} < \theta_n^{(1)} < 1.$$

Applying the corollary of Lemma 3 to each pair of the polynomials $\Phi_0'(x), \Phi_1'(x), \cdots, \Phi_n'(x), R_{n+1}'(x)$, we see that there are similar inequalities for the roots of the kth derivatives $\Phi_0^{(k)}(x), \Phi_1^{(k)}(x), \cdots, \Phi_n^{(k)}(x), R_{n+1}^{(k)}(x)$:

$$0 < \theta_1^{(k)} < \beta_1^{(k)} < \xi_{n-1,k}^{(1)} < \xi_{n-2,k}^{(1)} < \cdots < \xi_{1,k}^{(1)} < \alpha_2^{(k)} < \theta_2^{(k)} < \beta_2^{(k)} < \cdots$$
$$\cdots < \alpha_i^{(k)} < \theta_i^{(k)} < \beta_i^{(k)} < \xi_{n-1,k}^{(i)} < \xi_{n-2,k}^{(i)} < \cdots$$
$$\cdots < \xi_{1,k}^{(i)} < \alpha_{i+1}^{(k)} < \theta_{i+1}^{(k)} < \beta_{i+1}^{(k)} < \cdots$$
$$\cdots < \alpha_{n-k}^{(k)} < \theta_{n-k}^{(k)} < \beta_{n-k}^{(k)} < \xi_{n-1,k}^{(n-k)} < \xi_{n-2,k}^{(n-k)} < \cdots$$
$$\cdots < \xi_{1,k}^{(n-k)} < \alpha_{n-k+1}^{(k)} < \theta_{n-k+1}^{(k)} < 1.$$

Therefore inside each interval of the form

$$[\alpha_i^{(k)}, \beta_i^{(k)}] \quad (k = 1, 2, \cdots, n-1;\ i = 1, 2, \cdots, n-k+1)$$

there is a single root of $R_{n+1}^{(k)}(x)$, and each of the derivatives $\Phi_0^{(k)}(x)$, $\Phi_1^{(k)}(x), \cdots, \Phi_n^{(k)}(x)$ preserves its sign. Applying Lemma 2 to the polynomials $R_{n+1}(x)$, $\Phi_0(x), \cdots, \Phi_n(x)$, we reach the conclusion that all the numbers $\Phi_0^{(k)}(\theta_i^{(k)}), \Phi_1^{(k)}(\theta_i^{(k)}), \cdots, \Phi_n^{(k)}(\theta_i^{(k)})$ have the same sign (plus for $i = n-k+1, n-k-1, n-k-3, \cdots$, and minus for $i = n-k$, $n-k-2, n-k-4, \cdots$). It follows that if $\xi \in [\alpha_i^{(k)}, \beta_i^{(k)}]$, all the numbers $\Phi_0^{(k)}(\xi), \Phi_1^{(k)}(\xi), \cdots, \Phi_n^{(k)}(\xi)$ have the same sign, and in addition only $\Phi_0^{(k)}(\xi) = 0$ at $\xi = \alpha_i^{(k)}$ $(i = 1, 2, \cdots, n-k+1)$, while only $\Phi_n^{(k)}(\xi) = 0$ at $\xi = \beta_i^{(k)}$ $(i = 1, 2, \cdots, n-k)$. Turning to formula (5), we see that $\delta_0, \delta_1, \cdots, \delta_n$ have alternating signs, and condition 2) of the test is satisfied if $\xi \in [\alpha_i^{(k)}, \beta_i^{(k)}]$. At the left-hand endpoints $(\alpha_i^{(k)})_{i=2}^{n-k+1}$ loses the weight δ_0 at the node $\tau_0 = 0$, i.e. $\delta_0 = 0$, and at the right-hand endpoints $(\beta_i^{(k)})_{i=1}^{n-k}$ loses the weight δ_n at the node $\tau_n = 1$, i.e. $\delta_n = 0$. Consequently service of the functional $F_\xi^{(k)}$ by the polynomials $\pm T_n(x)$ ends (or begins) at these points (cf. [**4**], [**5**]).

Thus in the intervals $[\alpha_{n-k+1}^{(k)}, 1], [\alpha_{n-k-1}^{(k)}, \beta_{n-k-1}^{(k)}], \cdots$ the functional $F_\xi^{(k)}$ is served by $+T_n(x)$; and in the intervals

$$[\alpha_{n-k}^{(k)}, \beta_{n-k}^{(k)}], [\alpha_{n-k-2}^{(k)}, \beta_{n-k-2}^{(k)}], \cdots,$$

by $-T_n(x)$. When $k = n$ formula (5) takes the form

$$\delta_j = \frac{(-1)^{n-j} n!}{\prod_{i \neq j} |\tau_j - \tau_i|} \quad (j = 0, 1, \cdots, n),$$

and $F_\xi^{(n)}$ is served by $+T_n(x)$ for $\xi \in [0,1]$. This completes the proof of Theorem 1.

THEOREM 2. *Between the Čebyšev intervals there are open Zolotarev intervals* $(\beta_i^{(k)}, \alpha_{i+1}^{(k)})$ $(k = 1, 2, \cdots, n-1;\ i = 1, 2, \cdots, n-k)$ *at whose points the functional* (3) *is served by all the polynomials of passport* $[n, n, 0]$[3] *(denoted by* $Q_n(x, \vartheta)$*) and only by these, and indeed by each one at that point of each interval where*

$$\left[\frac{\partial^k R_n(x, \vartheta)}{\partial x^k}\right]_{\substack{x=\xi \\ \vartheta = \vartheta_\xi}} = 0.$$

Here $R_n(x, \vartheta) = \prod_{i=1}^n (x - \sigma_i)$ *is the resolvent of* $Q_n(x, \vartheta)$, $(\overset{\pm}{\sigma_i})_1^n$ *is its distribution, and* ϑ *is the variable leading coefficient of* $Q_n(x, \vartheta)$ $(-2^{2n-1} < \vartheta < 2^{2n-1})$.

PROOF. Let $L_n(x)$ be any polynomial of passport $[n, n, 0]$, let $(\overset{\pm}{\sigma_i})_1^n$ be its distribution, let

$$R_n(x) = \prod_{i=1}^n (x - \sigma_i) = \sum_{l=0}^n r_l x^l$$

be its resolvent, and let ξ_0 be a root of $R_n^{(k)}(x) = 0$. We shall show that $L_n(x)$ serves the functional $F_{\xi_0}^{(k)}$. The first n equations of the system (V) take the form

$$\sum_{i=1}^n \delta_i \sigma_i^l = 0 \qquad (l = 0, 1, \cdots, k-1),$$

$$\sum_{i=1}^n \delta_i \sigma_i^l = \frac{l!}{(l-k)!} \xi_0^{l-k} \quad (l = k, k+1, \cdots, n-1),$$

and the solution is given by

$$\delta_j = \frac{(-1)^{n-j-1}}{\prod_{i \neq j} |\sigma_j - \sigma_i|} \left[\frac{d^k}{dx^k} \left(\frac{R_n(x)}{x - \sigma_j}\right)\right]_{x=\xi_0} \quad (j = 1, 2, \cdots, n).$$

Condition 2) of the test for extremal character is satisfied since, by Lemma 2, the numbers

---

3) See [6].

$$\left[\frac{d^k}{dx^k}\left(\frac{R_n(x)}{x-\sigma_1}\right)\right]_{x=\xi_0}, \left[\frac{d^k}{dx^k}\left(\frac{R_n(x)}{x-\sigma_2}\right)\right]_{x=\xi_0}, \cdots, \left[\frac{d^k}{dx^k}\left(\frac{R_n(x)}{x-\sigma_n}\right)\right]_{x=\xi_0}$$

have the same sign. Condition 1), that

$$\frac{n!}{(n-k)!}\xi_0^{n-k} = \sum_{i=1}^{n} \delta_i \sigma_i^n,$$

i.e. that the entire system (V) is consistent, yields the equation

$$F_\xi^{(k)}[R_n(x)] = R_n^{(k)}(\xi) = \sum_{l=0}^{n} r_l \sum_{i=1}^{n} \delta_i \sigma_i^l = \sum_{i=1}^{n} \delta_i R_n(\sigma_i) = 0,$$

i.e. $R_n^{(k)}(\xi) = 0$, which is satisfied at ξ_0 and nowhere else.

Thus each polynomial $L_n(x)$ serves $F_\xi^{(k)}$ at $n-k$ points, the roots of $R_n^{(k)}(x) = 0$, lying one in each of the intervals $(\beta_i^{(k)}, \alpha_{i+1}^{(k)})$ $(i = 1, 2, \cdots, n-k)$.

If we delete from the set $\{Q_n(x, \vartheta)\}$ the Čebyšev transformations $\pm T_n(\nu x)$ and $\pm T_n(\nu \cdot \overline{1-x})$ that the set contains, the remaining polynomials are, up to constant factors, exactly the polynomials of E. I. Zolotarev [8], which he expressed in terms of elliptic functions; we denote them by $Z_n(x, \vartheta)$.

At the endpoints $(\beta_i^{(k)})_{i=1}^{n-k}$ and $(\alpha_i^{(k)})_{i=2}^{n-k+1}$ the functional $F_\xi^{(k)}$ loses its weight at, respectively, the nodes $\tau_n = 1$ $(\delta_n = 0)$ or $\tau_0 = 0$ $(\delta_0 = 0)$; hence in $(\beta_i^{(k)}, \alpha_{i+1}^{(k)})$, as ξ moves from $\beta_i^{(k)}$ to $\alpha_{i+1}^{(k)}$ (assuming, for definiteness, that $F_\xi^{(k)}$ is served in $[\alpha_i^{(k)}, \beta_i^{(k)}]$ by $+T_n(x)$), service changes in the following order:

$$+T_n(\nu x)\left(\cos^2\frac{\pi}{2n} \leqq \nu < 1\right), \quad +Z_n(x, \vartheta)\left(0 < \vartheta < 2^{2n-1}\cos^{2n}\frac{\pi}{2n}\right),$$

$$-T_{n-1}(x), \quad (-1)^{n-1}Z_n(1-x, \vartheta), \quad (-1)^{n-1}T_n(\nu \cdot \overline{1-x})$$

without a break, according to the theorem on continuous deformation (cf. [**4**], [**5**]), and without repetition until the extremal polynomial becomes $-T_n(x)$ at $\alpha_{i+1}^{(k)}$.

This completes the proof of Theorem 2.

Therefore if E_T is the set of points of Čebyšev intervals and E_z is its complement with respect to $[0,1]$, we have the following inequalities for reduced polynomials:

$$|P_n^{(k)}(\xi)| \leqq \begin{cases} |T_n^{(k)}(\xi)| = N_k(\xi) & \text{for } \xi \in E_T, \\ |Q_n^{(k)}(\xi, \vartheta_\xi)| = N_k(\xi) & \text{for } \xi \in E_z, \end{cases}$$

where ϑ_ξ and ξ are connected, by Theorem 2, by the equation

$$(6) \qquad \frac{\partial^k R_n(\xi, \vartheta)}{\partial \xi^k} = 0.$$

The norm $N_k(\xi)$ is a continuous function of ξ.

THEOREM 3. *In each interior Čebyšev interval the norm $N_k(\xi)$ takes its maximum value*

$$\max_{[\alpha_i^{(k)}, \beta_i^{(k)}]} N_k(\xi) = N_k(\gamma_i^{(k)})$$

just once, where $(\gamma_i^{(k)})_{i=2}^{n-k}$ are the roots of $T_n^{(k+1)}(x) = 0$; in the two boundary Čebyšev intervals the norm $N_k(\xi)$ decreases monotonically from the endpoints of $[0,1]$ inward.

PROOF. Theorem 3 of Voronovskaja [7] shows that the roots $(\gamma_i^{(1)})_{i=2}^{n-1}$ of $T_n''(x)$ lie one each in the interior intervals $[\alpha_i^{(1)}, \beta_i^{(1)}]$ $(i = 2, 3, \cdots, n-1)$. Applying the corollary of V. A. Markov's Lemma 3 to each pair of the polynomials $\Phi_0'(x)$, $T_n''(x)$, $\Phi_n'(x)$, we see that for each $k \leqq n-2$ the roots $(\gamma_i^{(k)})_{i=2}^{n-k}$ of $T_n^{(k+1)}(x)$ lie one each in the interior intervals $[\alpha_i^{(k)}, \beta_i^{(k)}]$ $(i = 2, 3, \cdots, n-k)$. It follows that the norm $N_k(\xi)$ attains its maximum once in each interior Čebyšev interval, and decreases monotonically, from the endpoints of $[0,1]$ inward, in the boundary Čebyšev intervals, q.e.d.

REMARK. If $k \equiv n \pmod 2$, there is a Čebyšev interval

$$[\alpha_{(n-k+2)/2}^{(k)}, 1 - \alpha_{(n-k+2)/2}^{(k)}],$$

containing the point $\xi = \gamma_{(n-k+2)/2}^{(k)} = \frac{1}{2}$, and

$$N_k\left(\frac{1}{2}\right) = \left| T_n^{(k)}\left(\frac{1}{2}\right) \right| = \begin{cases} 2^k n(n^2 - 1^2) \cdots (n^2 - \overline{k-2}^2) & \text{for } n \text{ odd}, \\ 2^k n^2(n^2 - 2^2) \cdots (n^2 - \overline{k-2}^2) & \text{for } n \text{ even}. \end{cases}$$

We turn to the norm $N_k(\xi)$ on the Zolotarev intervals. In each interval $(\beta_i^{(k)}, \alpha_{i+1}^{(k)})$ there is a unique point $\xi_{i,k}^*$ $(i = 1, 2, \cdots, n-k)$ at which the functional $F_\xi^{(k)}$ is served by one of the polynomials $\pm T_{n-1}(x)$. The points $(\xi_{i,k}^*)_{i=1}^{n-k}$ can be found as the roots of $R_n^{(k)}(x) = 0$, where $R_n(x)$ is the resolvent of $T_{n-1}(x)$. We call $(\beta_i^{(k)}, \xi_{i,k}^*)$ and $(\xi_{i,k}^*, \alpha_{i+1}^{(k)})$ the left-hand and right-hand parts of the interval $(\beta_i^{(k)}, \alpha_{i+1}^{(k)})$.

THEOREM 4. *In the interval $(\beta_i^{(k)}, \alpha_{i+1}^{(k)})$ $(i = 1, 2, \cdots, n-k)$ the norm $N_k(\xi)$ varies monotonically at each point ξ at which the $(k+1)$th derivative of the extremal polynomial is not zero.*

PROOF. Suppose for definiteness that the extremal polynomial at $\xi = \beta_i^{(k)}$ is $+T_n(x)$. Then in some interval $(\beta_i^{(k)}, A_i^{(k)})$, $\beta_i^{(k)} < A_i^{(k)} < \xi_{i,k}^*$, the Čebyšev transformations $T_n(\nu x)$ will be extremal (cf. Theorem 2). We have

$$N_k(\xi) = T_n^{(k)}(\nu\xi) \cdot \nu^k$$

and

$$N_k'(\xi) = -kT_n^{(k)}(\nu\xi) \cdot \frac{[\beta_i^{(k)}]^k}{\xi^{k+1}};$$

in fact, the quantity $T_n^{(k)}(\nu\xi) \cdot \nu^k$ must be maximized over ν, i.e.

$$T_n^{(k+1)}(\nu\xi) \cdot \xi\nu^k + T_n^{(k)}(\nu\xi) k\nu^{k-1} = 0,$$

or

$$T_n^{(k+1)}(\nu\xi) \cdot \nu\xi + kT_n^{(k)}(\nu\xi) = 0,$$

whence it follows that $\nu\xi = \text{const}$. Therefore $\nu = \beta_i^{(k)}/\xi$, $A_i^{(k)} = \beta_i^{(k)}/\cos^2(\pi/2n)$ ($\xi = \beta_i^{(k)}$ for $\nu = 1$; $\xi = A_i^{(k)}$ for $\nu = \cos^2(\pi/2n)$); $T_n^{(k)}(\nu\xi) > 0$, $N_k'(\xi) < 0$, and the norm $N_k(\xi)$ decreases in $(\beta_i^{(k)}, A_i^{(k)})$. The monotone character of the norm is proved similarly for the other Čebyšev transformations.

Thus it remains to prove the theorem for the part of $(\beta_i^{(k)}, \alpha_{i+1}^{(k)})$ in which the polynomials $Z_n(x, \vartheta)$ are extremal. Let $A_i^{(k)} < \xi_1 < \xi_{i,k}^*$; then $Z_n(x, \vartheta_{\xi_1})$ serves $F_{\xi_1}^{(k)}$, and

$$N_k(\xi_1) = [Z_n^{(k)}(x, \vartheta_{\xi_1})]_{x=\xi_1},$$
$$[R_n^{(k)}(x, \vartheta_{\xi_1})]_{x=\xi_1} = 0,$$

where $R_n(x, \vartheta_{\xi_1})$ is the resolvent of $Z_n(x, \vartheta_{\xi_1})$ (cf. Theorem 2). We have

$$N_k'(\xi) = \frac{\partial^{k+1} Z_n(\xi, \vartheta)}{\partial \xi^{k+1}} + \frac{\partial^{k+1} Z_n(\xi, \vartheta)}{\partial \xi^k \partial \vartheta} \frac{d\vartheta}{d\xi}.$$

Using the fundamental relation connecting a Zolotarev polynomial and its resolvent,

$$\frac{\partial Z_n(\xi, \vartheta)}{\partial \vartheta} = R_n(\xi, \vartheta)$$

(cf. [4], [6]), we obtain

$$\frac{\partial^{k+1} Z_n(\xi, \vartheta)}{\partial \xi^k \partial \vartheta} = \frac{\partial^k R_n(\xi, \vartheta)}{\partial \xi^k}.$$

consequently

$$N_k'(\xi_1) = \left(\frac{\partial^{k+1} Z_n(\xi, \vartheta)}{\partial \xi^{k+1}}\right)_{\substack{\xi=\xi_1 \\ \vartheta=\vartheta_{\xi_1}=\vartheta_1}} + \left(\frac{\partial^k R_n(\xi, \vartheta)}{\partial \xi^k} \cdot \frac{d\vartheta}{d\xi}\right)_{\substack{\xi=\xi_1 \\ \vartheta=\vartheta_{\xi_1}=\vartheta_1}} = Z_n^{(k+1)}(\xi_1 \vartheta_1).$$

Remark. The norm $N_k(\xi)$ has a continuous derivative $N_k'(\xi)$. In fact,

$$\lim_{\xi\to\beta_i^{(k)}-0} N_k'(\xi) = N_k'(\beta_i^{(k)}-0) = T_n^{(k+1)}(\beta_i^{(k)})$$

and

$$\lim_{\xi\to\beta_i^{(k)}+0} N_k'(\xi) = N_k'(\beta_i^{(k)}+0) = -\frac{k}{\beta_i^{(k)}} T_n^{(k)}(\beta_i^{(k)}).$$

Since

$$A_i^{(k)} = \frac{\beta_i^{(k)}}{\cos^2(\pi/2n)}$$

and

$$Z_n\left(x, 2^{2n-1}\cos^{2n}\frac{\pi}{2n}\right) = T_n\left(x\cos^2\frac{\pi}{2n}\right),$$

then

$$\lim_{\xi\to A_i^{(k)}+0} N_k'(\xi) = N_k'(A_i^{(k)}+0) = \cos^{2(k+1)}\frac{\pi}{2n}\cdot T_n^{(k+1)}(\beta_i^{(k)})$$

and

$$\lim_{\xi\to A_i^{(k)}-0} N_k'(\xi) = N_k'(A_i^{(k)}-0) = -kT_n^{(k)}(\beta_i^{(k)})\frac{\cos^{2(k+1)}(\pi/2n)}{\beta_i^{(k)}}.$$

Using the relation

$$T_n^{(k+1)}(\beta_i^{(k)})\beta_i^{(k)} + kT_n^{(k)}(\beta_i^{(k)}) = 0$$

(cf. Theorem 4), we see that

$$N_k'(\beta_i^{(k)}-0) = N_k'(\beta_i^{(k)}+0)$$
$$N_k'(A_i^{(k)}-0) = N_k'(A_i^{(k)}+0).$$

By the symmetry of the norm (cf. formula (4)) we infer the continuity of $N_k'(\xi)$ at the points $(\alpha_i^{(k)})_{i=2}^{n-k+1}$ and $(1-A_i^{(k)})_{i=1}^{n-k}$. At the remaining points $N_k'(\xi)$ is obviously continuous.

We now find an expression for the second derivative $N_k''(\xi)$ in the interval $(A_i^{(k)}, \xi_{i,k}^*)$. Since

$$N_k'(\xi) = Z_n^{(k+1)}(\xi, \vartheta_\xi),$$

we have

$$N_k''(\xi) = \frac{\partial Z_n^{(k+1)}(\xi,\vartheta_\xi)}{\partial\xi} + \frac{\partial Z_n^{(k+1)}(\xi,\vartheta_\xi)}{\partial\vartheta}\cdot\frac{d\vartheta}{d\xi}$$
$$= Z_n^{(k+2)}(\xi,\vartheta_\xi) + R_n^{(k+1)}(\xi,\vartheta_\xi)\cdot\frac{d\vartheta}{d\xi}.$$

Differentiating (6) with respect to ξ, we obtain

$$R_n^{(k+1)}(\xi, \vartheta) + \frac{\partial R_n^{(k)}(\xi, \vartheta)}{\partial \vartheta} \cdot \frac{d\vartheta}{d\xi} = 0,$$

whence we find

$$\frac{d\vartheta}{d\xi} = -\frac{R_n^{(k+1)}(\xi, \vartheta)}{\partial R_n^{(k)}(\xi, \vartheta)/\partial \vartheta}.$$

We now compute $\partial R_n(x, \vartheta)/\partial \vartheta$. Since

$$R_n(x, \vartheta) = \prod_{i=1}^{n} [x - \sigma_i(\vartheta)],$$

we have

$$\frac{\partial R_n(x, \vartheta)}{\partial \vartheta} = -\sum_{i=1}^{n} \frac{d\sigma_i}{d\vartheta} \cdot R_i(x, \vartheta),$$

where

$$R_i(x, \vartheta) = \frac{R_n(x, \vartheta)}{x - \sigma_i}.$$

In addition, the following relation holds between $Z_n(x, \vartheta)$ and its resolvent (cf. [**6**]):

$$n\vartheta(x - \lambda) R_n(x, \vartheta) = x(x-1) Z_n'(x, \vartheta), \tag{7}$$

where λ is the zero of $Z_n'(x, \vartheta)$ outside $[0, 1]$.

Differentiating (7) with respect to ϑ, we obtain

$$n(x - \lambda) R_n(x, \vartheta) - n\vartheta R_n(x, \vartheta) \frac{d\lambda}{d\vartheta} + n\vartheta(x - \lambda) \frac{\partial R_n(x, \vartheta)}{\partial \vartheta}$$

$$= x(x-1) \frac{\partial Z_n'(x, \vartheta)}{\partial \vartheta} = x(x-1) R_n'(x, \vartheta) = x(x-1) \sum_{i=1}^{n} R_i(x, \vartheta).$$

Putting $x = \sigma_i$ in the extreme parts of the preceding equation, we find

$$n\vartheta(\sigma_i - \lambda) \frac{\partial R_n(\sigma_i, \vartheta)}{\partial \vartheta} = \sigma_i(\sigma_i - 1) R_i(\sigma_i, \vartheta).$$

But

$$\frac{\partial R_n(\sigma_i, \vartheta)}{\partial \vartheta} = -\frac{d\sigma_i}{d\vartheta} \cdot R_i(\sigma_i, \vartheta),$$

hence

This completes the proof of Theorem 4.

$$\frac{d\sigma_i}{d\vartheta} = \frac{\sigma_i(1-\sigma_i)}{n\vartheta(\sigma_i-\lambda)}$$

and

$$\frac{\partial R_n(x,\vartheta)}{\partial\vartheta} = \sum_{i=1}^{n} \frac{\sigma_i(\sigma_i-1)}{n\vartheta(\sigma_i-\lambda)} R_i(x,\vartheta).$$

Thus

$$\frac{d\vartheta}{d\xi} = \frac{n\vartheta R_n^{(k+1)}(\xi,\vartheta)}{\sum_{i=1}^{n} \frac{\sigma_i(1-\sigma_i)}{\sigma_i-\lambda} R_i^{(k)}(\xi,\vartheta)}.$$

We introduce the notation

$$\phi(x,\vartheta) = \sum_{i=1}^{n} \frac{\sigma_i(\sigma_i-1)}{\sigma_i-\lambda} R_i(x,\vartheta)$$

and

$$\psi(x,\vartheta) = \sum_{i=1}^{n} \frac{R_i(x,\vartheta)}{\sigma_i-\lambda};$$

then we have

$$\phi(x,\vartheta) = \sum_{i=1}^{n} (\sigma_i - 1 + \lambda) R_i(x,\vartheta) + \lambda(\lambda-1)\psi(x,\vartheta)$$

$$(8) \qquad = -\sum_{i=1}^{n}(x-\sigma_i)R_i(x,\vartheta) + (x-1+\lambda)\sum_{i=1}^{n} R_i(x,\vartheta) + \lambda(\lambda-1)\psi(x,\vartheta)$$

$$= -nR_n(x,\vartheta) + (x-1+\lambda)R_n'(x,\vartheta) + \lambda(\lambda-1)\psi(x,\vartheta).$$

Put

$$\chi(x,\vartheta) = (x-\lambda)\psi(x,\vartheta);$$

then

$$\chi(\sigma_i,\vartheta) = (\sigma_i-\lambda)\psi(\sigma_i,\vartheta) = R_i(\sigma_i,\vartheta), \quad \chi(\lambda,\vartheta) = 0,$$

whence it is clear that

$$(9) \qquad \chi(x,\vartheta) = (x-\lambda)\psi(x,\vartheta) = R_n'(x,\vartheta) - \frac{R_n'(\lambda,\vartheta)}{R_n(\lambda,\vartheta)} R_n(x,\vartheta).$$

THEOREM 5. *In each Zolotarev interval the norm $N_k(\xi)$ has just one minimum, at the point $\xi = \xi_{0,i}^{(k)}$ at which $Z_n^{(k+1)}(\xi,\vartheta_\xi) = 0$, and*

$$N_k(\xi_{0,i}^{(k)}) = \min_{(\beta_i^{(k)},\alpha_{i+1}^{(k)})} N_k(\xi) \leqq |T_{n-1}^{(k)}(\xi_{i,k}^*)|$$

$$(i = 1, 2, \cdots, n-k;\ k = 1, 2, \cdots, n-1).$$

If $\beta_i^{(k)} > \frac{1}{2}$ *we have* $\beta_i^{(k)} < \xi_{0,i}^{(k)} < \xi_{i,k}^*$; *if* $\alpha_{i+1}^{(k)} < \frac{1}{2}$ *we have* $\xi_{i,k}^* < \xi_{0,i}^{(k)} < \alpha_{i+1}^{(k)}$.

PROOF. Let $Z_n^{(k+1)}(\xi_0, \vartheta_{\xi_0}) = 0$, where $\beta_i^{(k)} < \xi_0 < \xi_{i,k}^*$. We differentiate (7) $k+1$ times with respect to x, and then put $x = \xi_0$ and $\vartheta = \vartheta_{\xi_0} = \vartheta_0$. Taking into account that $R_n^{(k)}(\xi_0, \vartheta_0) = 0$ and $Z_n^{(k+1)}(\xi_0, \vartheta_0) = 0$, we obtain

$$n\vartheta_0(\xi_0 - \lambda) R_n^{(k+1)}(\xi_0, \vartheta_0) = \xi_0(\xi_0 - 1) Z_n^{(k)}(\xi_0, \vartheta_0) + (k+1) k Z_n^{(k)}(\xi_0, \vartheta_0). \tag{10}$$

Now differentiate (8) and (9) k times with respect to x, and then put $x = \xi_0$ and $\vartheta = \vartheta_0$; we obtain

$$\chi^{(k)}(\xi_0, \vartheta_0) = (\xi_0 - \lambda)\psi^{(k)}(\xi_0, \vartheta_0) + k\psi^{(k-1)}(\xi_0, \vartheta_0) = R_n^{(k-1)}(\xi_0, \vartheta_0),$$

whence

$$\psi^{(k)}(\xi_0, \vartheta_0) = \frac{R_n^{(k+1)}(\xi_0, \vartheta_0) - k\psi^{(k-1)}(\xi_0, \vartheta_0)}{\xi_0 - \lambda}$$

and therefore

$$\begin{aligned}\phi^{(k)}(\xi_0, \vartheta_0) &= (\xi_0 - 1 + \lambda) R_n^{(k+1)}(\xi_0, \vartheta_0) \\ &\quad + \lambda(\lambda - 1)\frac{R_n^{(k+1)}(\xi_0, \vartheta_0) - k\psi^{(k-1)}(\xi_0, \vartheta_0)}{\xi_0 - \lambda} \\ &= \frac{\xi_0(\xi_0 - 1)}{\xi_0 - \lambda} R_n^{(k+1)}(\xi_0, \vartheta_0) - \frac{k\lambda(\lambda - 1)}{\xi_0 - \lambda}\psi^{(k-1)}(\xi_0, \vartheta_0).\end{aligned} \tag{11}$$

Since

$$\begin{aligned}N_k''(\xi_0) &= Z_n^{(k+2)}(\xi_0, \vartheta_0) + R_n^{(k+1)}(\xi_0, \vartheta_0)\left(\frac{d\vartheta}{d\xi}\right)_{\xi=\xi_0} \\ &= Z_n^{(k+2)}(\xi_0, \vartheta_0) - \frac{n\vartheta_0 R_n^{(k+1)}(\xi_0, \vartheta_0)}{\dfrac{\phi^{(k)}(\xi_0, \vartheta_0)}{R_n^{(k+1)}(\xi_0, \vartheta_0)}},\end{aligned}$$

we find, by using (10) and (11), that

$$N_k''(\xi_0) = \frac{kZ_n^{(k)}(\xi_0, \vartheta_0)}{\lambda - \xi_0} \cdot \frac{k + 1 + \lambda(\lambda - 1)\dfrac{\psi^{(k-1)}(\xi_0, \vartheta_0)}{R_n^{(k+1)}(\xi_0, \vartheta_0)} \cdot \dfrac{Z_n^{(k+2)}(\xi_0, \vartheta_0)}{Z_n^{(k)}(\xi_0, \vartheta_0)}}{\dfrac{\xi_0(\xi_0 - 1)}{\xi_0 - \lambda} - \dfrac{k\lambda(\lambda - 1)}{\xi_0 - \lambda} \cdot \dfrac{\psi^{(k-1)}(\xi_0, \vartheta_0)}{R_n^{(k+1)}(\xi_0, \vartheta_0)}}. \tag{12}$$

This is similar to V. A. Markov's formula (118) [**1**]. Now we have

$$\frac{\xi_0(\xi_0 - 1)}{\xi_0 - \lambda} - \frac{k\lambda(\lambda - 1)}{\xi_0 - \lambda} \cdot \frac{\psi^{(k-1)}(\xi_0, \vartheta_0)}{R_n^{(k+1)}(\xi_0, \vartheta_0)} = \frac{\phi^{(k)}(\xi_0, \vartheta_0)}{R_n^{(k+1)}(\xi_0, \vartheta_0)} = -\frac{n\vartheta_0}{(d\vartheta/d\xi)_{\xi=\xi_0}} > 0,$$

since as ξ increases from $A_i^{(k)}$ to $\xi_{i,k}^*$ the parameter ϑ decreases monotonically and $d\vartheta/d\xi < 0$; in addition,

$$\frac{Z_n^{(k+2)}(\xi_0, \vartheta_0)}{Z_n^{(k)}(\xi_0, \vartheta_0)} < 0,$$

by Markov's Lemma 1, and

$$\frac{\psi^{(k-1)}(\xi_0, \vartheta_0)}{R_n^{(k+1)}(\xi_0, \vartheta_0)} < 0,$$

by Markov's Lemma 3. Thus $N_k''(\xi_0)$ and $Z_n^{(k)}(\xi_0, \vartheta_0) = N_k(\xi_0)$ have the same sign. Consequently if we have

$$Z_n^{(k+1)}(\xi_0, \vartheta_0) = 0,$$

then

$$N_k(\xi_0) = \min_{(\beta_i^{(k)}, \alpha_{i+1}^{(k)})} N_k(\xi),$$

and there can be no more than one such point ξ_0 in each Zolotarev interval. We now show that there exists one such point in each $(\beta_i^{(k)}, \alpha_{i+1}^{(k)})$. We have

$$(n-1)2^{2n-3}R_n(x) = x(x-1)T'_{n-1}(x),$$

where $R_n(x)$ is the resultant of $T_{n-1}(x)$. Differentiating the preceding equation k times, we obtain

$$(n-1)2^{2n-3}R_n^{(k)}(x) = x(x-1)T_{n-1}^{(k+1)}(x) + k(2x-1)T_{n-1}^{(k)}(x) + k(k+1T_{n-1}^{(k-1)}(x).$$

In addition,

$$x(1-x)T_{n-1}^{(k+1)}(x) - (2k-1)(x-\tfrac{1}{2})T_{n-1}^{(k)}(x) + (\overline{n-1}^2 - \overline{k-1}^2)T_{n-1}^{(k-1)}(x) = 0.$$

In the last two equations put $x = \xi_{i,k}^*$, remember that $R_n^{(k)}(\xi_{i,k}^*) = 0$, and then eliminate $T_{n-1}^{(k-1)}(\xi_{i,k}^*)$ to obtain

$$\xi_{i,k}^*(1-\xi_{i,k}^*)[\overline{n-1}^2 - \overline{k-1}^2 + k(k+1)]T_{n-1}^{(k+1)}(\xi_{i,k}^*) = k(2\xi_{i,k}^* - 1)[\overline{n-1}^2 - \overline{k-1}^2 + (k-\tfrac{1}{2})(k+1)]T_{n-1}^{(k)}(\xi_{i,k}^*).$$

If $\xi_{i,k}^* > \frac{1}{2}$ in this equation, i.e. $\beta_i^{(k)} > \frac{1}{2}$, we have

$$\operatorname{sgn} T_{n-1}^{(k+1)}(\xi_{i,k}^*) = \operatorname{sgn} T_{n-1}^{(k)}(\xi_{i,k}^*)$$

and $N_k'(\xi_{i,k}^*) > 0$. On the other hand,

$$N_k'(A_i^{(k)}) = -k|T_n^{(k)}(\beta_i^{(k)})|\frac{\cos^{2(k+1)}(\pi/2n)}{\beta_i^{(k)}} < 0.$$

Consequently there is a unique point $\xi_{0,i}^{(k)}$ in the interval $(A_i^{(k)}, \xi_{i,k}^*)$ such that

$$N_k'(\xi_{0,i}^{(k)}) = 0.$$

By (4), this point $\xi_{0,i}^{(k)}$ lies in the interval $(\xi_{i,k}^*, \alpha_{i+1}^{(k)})$ if $\alpha_{i+1}^{(k)} < \frac{1}{2}$. This completes the proof of Theorem 5.

REMARK 1. If $k+1 \equiv n \pmod 2$ there is a Zolotarev interval

$$(\beta_{(n-k+1)/2}^{(k)}, 1 - \beta_{(n-k+1)/2}^{(k)})$$

containing the point

$$\xi = \xi_{0,(n-k+1)/2}^{(k)} = \xi_{(n-k+1)/2,k}^* = \tfrac{1}{2},$$

and

$$N_k(\tfrac{1}{2}) = |T_{n-1}^{(k)}(\tfrac{1}{2})|$$
$$= \begin{cases} 2^k(n-1)(\overline{n-1}^2 - 1^2)\cdots(\overline{n-1}^2 - \overline{k-2}^2) & \text{for } n \text{ even},\\ 2^k(n-1)^2(\overline{n-1}^2 - 2^2)\cdots(\overline{n-1}^2 - \overline{k-2}^2 & \text{for } n \text{ odd}. \end{cases}$$

REMARK 2. Since $T_n(x) = \cos n\theta$, where $\theta = \arccos(2x-1)$, we have

$$T_n'(x) = -n\sin n\theta \frac{d\theta}{dx} = \frac{2n \sin n\theta}{\sin\theta} = 4n[\cos(n-1)\theta + \cos(n-3)\theta + \cdots]$$

and

$$T_n^{(k)}(x) = \sum_{i=0}^{n-k} a_{k,i}\cos i\theta,$$

with $a_{k,i} \geqq 0$ (cf. [2]). It follows that

$$\max_{[0,1]} N_k(\xi) = T_n^{(k)}(1),$$

and we obtain V. A. Markov's inequality (2′).

THEOREM 6. *The discontinuities of the second derivative* $N_k''(\xi)$ *of the norm are*

$$(\alpha_i^{(k)})_{i=2}^{n-k+1}, \quad (\beta_i^{(k)})_{i=1}^{n-k}, \quad (A_i^{(k)})_{i=1}^{n-k} \quad \textit{and} \quad (1 - A_i^{(k)})_{i=1}^{n-k}$$

PROOF. We have

$$N_k''(\beta_i^{(k)} - 0) = T_n^{(k+2)}(\beta_i^{(k)}),$$

$$N_k''(\beta_i^{(k)} + 0) = \frac{k(k+1)}{[\beta_i^{(k)}]^2} T_n^{(k)}(\beta_i^{(k)}) = -\frac{k+1}{\beta_i^{(k)}} T_n^{(k+1)}(\beta_i^{(k)})$$

and consequently

$$N_k''(\beta_i^{(k)}-0)-N_k''(\beta_i^{(k)}+0)=T_n^{(k+2)}(\beta_i^{(k)})+\frac{k+1}{\beta_i^{(k)}}T_n^{(k+1)}(\beta_i^{(k)})$$
$$=\frac{n2^{2n-1}}{\beta_i^{(k)}}\cdot\Phi_n^{(k+1)}(\beta_i^{(k)})\neq 0,$$
$$(n2^{2n-1}\Phi_n(x)=xT_n'(x)).$$

We have $\Phi_n^{(k+1)}(\beta_i^{(k)})>0$ for $i=n-k,\ n-k-2,\ n-k-4,\cdots$, and $\Phi_n^{(k+1)}(\beta_i^{(k)})<0$ for $i=n-k-1, n-k-3,\cdots$. Hence if we use the fact that the extremal polynomial is $+T_n(x)$ in the intervals $[\alpha_{n-k+1}^{(k)},1]$, $[\alpha_{n-k-1}^{(k)},\beta_{n-k-1}^{(k)}],\cdots$, and $-T_n(x)$ in $[\alpha_{n-k}^{(k)},\beta_{n-k}^{(k)}],[\alpha_{n-k-2}^{(k)},\beta_{n-k-2}^{(k)}],\cdots$ (cf. Theorem 1), we have

$$N_k''(\beta_i^{(k)}+0)>N_k''(\beta_i^{(k)}-0).$$

Using formula (4), we find that at the points $(\alpha_i^{(k)})_{i=2}^{n-k+1}$

$$N_k''(\alpha_i^{(k)}+0)<N_k''(\alpha_i^{(k)}-0).$$

In addition,

$$N_k''(A_i^{(k)}-0)=\frac{k(k+1)}{[\beta_i^{(k)}]^2}T_n^{(k)}(\beta_i^{(k)})\cos^{2(k+2)}\frac{\pi}{2n}$$
$$=-\frac{k+1}{\beta_i^{(k)}}T_n^{(k+1)}(\beta_i^{(k)})\cos^{2(k+2)}\frac{\pi}{2n}$$

and

$$N_k''(A_i^{(k)}+0)=\left[Z_n^{(k+2)}(\xi,\vartheta)+R_n^{(k+1)}(\xi,\vartheta)\cdot\frac{d\vartheta}{d\xi}\right]_{\substack{\xi=A_i^{(k)}\\ \vartheta=2^{2n-1}\cos^{2n}(\pi/2n)}}$$
$$=\cos^{2(k+2)}\frac{\pi}{2n}\cdot T_n^{(k+2)}(\beta_i^{(k)})+\frac{\Phi_n^{(k+1)}(\beta_i^{(k)})}{\cos^{2(n-k-1)}(\pi/2n)}\cdot\left(\frac{d\vartheta}{d\xi}\right)_{\xi=A_i^{(k)}},$$

since $A_i^{(k)}=\beta_i^{(k)}/\cos^2(\pi/2n)$ and

$$R_n\left(x,2^{2n-1}\cos^{2n}\frac{\pi}{2n}\right)=\prod_{i=1}^{n}\left(x-\frac{\tau_{i-1}}{\cos^2(\pi/2n)}\right)=\frac{\Phi_n(x\cos^2(\pi/2n))}{\cos^{2n}(\pi/2n)}$$

is the resolvent of the polynomial

$$Z_n\left(x,2^{2n-1}\cos^{2n}\frac{\pi}{2n}\right)=T_n\left(x\cos^2\frac{\pi}{2n}\right).$$

Consequently

$$\begin{aligned}&N_k''(A_i^{(k)}+0)-N_k''(A_i^{(k)}-0)\\ &\quad=\cos^{2(k+2)}\frac{\pi}{2n}\cdot\Phi_n^{(k+1)}(\beta_i^{(k)})\left[\frac{n2^{2n-1}}{\beta_i^{(k)}}+\frac{(d\vartheta/d\xi)_{\xi=A_i^{(k)}}}{\cos^{2(n+1)}(\pi/2n)}\right],\end{aligned}\tag{13}$$

where

$$\left(\frac{d\vartheta}{d\xi}\right)_{\xi=A_i^{(k)}} = -n\left[\frac{\vartheta R_n^{(k+1)}(\xi,\vartheta)}{\phi^{(k)}(\xi,\vartheta)}\right]_{\substack{\xi=A_i^{(k)}\\ \vartheta=2^{2n-1}\cos^{2n}(\pi/2n)}}$$

$$= -\frac{n2^{2n-1}\cos^{2(k+1)}(\pi/2n)\cdot\Phi_n^{(k+1)}(\beta_i^{(k)})}{\phi^{(k)}(A_i^{(k)},2^{2n-1}\cos^{2n}(\pi/2n))}.$$

Let us calculate $\phi^{(k)}(A_i^{(k)},2^{2n-1}\cos^{2n}(\pi/2n))$. Since

$$\phi\left(x,2^{2n-1}\cos^{2n}\frac{\pi}{2n}\right)$$

$$= -\frac{n\Phi_n\left(x\cos^2\frac{\pi}{2n}\right)}{\cos^{2n}\frac{\pi}{2n}} + \frac{x\Phi_n'\left(x\cos^2\frac{\pi}{2n}\right)}{\cos^{2(n-1)}\frac{\pi}{2n}} - \frac{R_n\left(x,2^{2n-1}\cos^{2n}\frac{\pi}{2n}\right)}{x-1}$$

(cf. formula (8)), if we differentiate k times and put $x=A_i^{(k)}$ we obtain

$$\phi^{(k)}\left(A_i^{(k)},2^{2n-1}\cos^{2n}\frac{\pi}{2n}\right) = \frac{\beta_i^{(k)}\Phi_n^{(k+1)}(\beta_i^{(k)})}{\cos^{2(n-k-1)}(\pi/2n)} - \Delta_n^{(k)}(A_i^{(k)}), \tag{14}$$

where

$$\Delta_n^{(k)}(A_i^{(k)}) = \left\{\frac{d}{dx^k}\left[\frac{R_n(x,2^{2n-1}\cos^{2n}(\pi/2n))}{x-1}\right]\right\}_{x=A_i^{(k)}}.$$

We now substitute the formula for $(d\vartheta/d\xi)_{\xi=A_i^{(k)}}$ into (13); then, using (14), we find after some calculation

$$N_k''(A_i^{(k)}+0) - N_k''(A_i^{(k)}-0)$$

$$= \left[\frac{\sin^2\frac{\pi}{2n}}{\cos^{2(n-k-1)}\frac{\pi}{2n}}\cdot\Phi_n^{(k+1)}(\beta_i^{(k)}) + \frac{\cos^2\frac{\pi}{2n}}{\beta_i^{(k)}}\cdot\Delta_n^{(k)}(A_i^{(k)})\right]\left(\frac{d\vartheta}{d\xi}\right)_{\xi=A_i^{(k)}} \neq 0,$$

since $\Phi_n^{(k+1)}(\beta_i^{(k)})$ and $\Delta_n^{(k)}(A_i^{(k)})$ have the same sign by Markov's Lemma 2.

Just as for the points $(\beta_i^{(k)})_{i=1}^{n-k}$ and $(\alpha_i^{(k)})_{i=2}^{n-k+1}$, we see that

$$N_k''(A_i^{(k)}+0) > N_k''(A_i^{(k)}-0)$$

and

$$N_k''(1-A_i^{(k)}+0) < N_k''(1-A_i^{(k)}-0).$$

At the remaining points $N_k''(\xi)$ is clearly continuous.

THEOREM 7. *The sum of the lengths of the Čebyšev intervals, or the measure of the set E_T, is k/n.*

PROOF. We have

$$\operatorname{mes} E_T = \sum_{i=1}^{n-k+1} (\beta_i^{(k)} - \alpha_i^{(k)}) = 1 + \sum_{i=1}^{n-k} \beta_i^{(k)} - \sum_{i=2}^{n-k+1} \alpha_i^{(k)}.$$

In addition,

$$\begin{aligned} \Phi_0^{(k)}(x) &= A_k \left(x^{n-k} - \sum_{i=2}^{n-k+1} \alpha_i^{(k)} \cdot x^{n-k-1} + \cdots \right), \\ \Phi_n^{(k)}(x) &= B_k \left(x^{n-k} - \sum_{i=1}^{n-k} \beta_i^{(k)} \cdot x^{n-k-1} + \cdots \right). \end{aligned} \tag{15}$$

On the other hand, since

$$\Phi_j(x) = x^n - \left(\frac{n+1}{2} - \tau_j \right) x^{n-1} + \cdots,$$

we have

$$\begin{aligned} \Phi_0^{(k)}(x) &= \frac{n!}{(n-k)!} x^{n-k} - \frac{n+1}{2} \frac{(n-1)!}{(n-k-1)!} x^{n-k-1} + \cdots, \\ \Phi_n^{(k)}(x) &= \frac{n!}{(n-k)!} x^{n-k} - \frac{n-1}{2} \frac{(n-1)!}{(n-k-1)!} x^{n-k-1} + \cdots. \end{aligned} \tag{16}$$

Comparing the expansions (15) and (16), we find that

$$\sum_{i=2}^{n-k+1} \alpha_i^{(k)} = \frac{n+1}{2} \left(1 - \frac{k}{n}\right), \quad \sum_{i=1}^{n-k} \beta_i^{(k)} = \frac{n-1}{2} \left(1 - \frac{k}{n}\right),$$

whence we obtain

$$\operatorname{mes} E_T = k/n.$$

It follows from Theorem 7 that as $n \to \infty$ with k fixed, the Čebyšev intervals contract to points, i.e. the Zolotarev polynomials displace the polynomials $T_n(x)$ on $[0,1]$. Conversely, if $n - k$ is fixed and $n \to \infty$, the polynomials $T_n(x)$ displace the Zolotarev polynomials.

In conclusion we compare the exact majorant $N_k(\xi)$ with Bernšteĭn's majorant [**9**]:

$$|P_n^{(k)}(\xi)| < \left[\frac{k}{\xi(1-\xi)} \right]^{k/2} \cdot n(n-1) \cdots (n-k+1) = B_k(\xi) \tag{17}$$

and Duffin and Schaeffer's majorant [**3**]:

$$|P_n^{(k)}(\xi)| \leqq |T_n^{(k)}(\xi) + iS_n^{(k)}(\xi)| \quad (S_n(x) = \sin n \arccos(2x-1)) \tag{18}$$

(we have written both inequalities for reduced polynomials on $[0,1]$; for $k = 1$ they are the same).

For $k = 2, 3, \cdots, n$ the majorant (17) never equals the exact majorant $N_k(\xi)$ (cf. [**9**], p. 27).

For large n, (17) was sharpened by Bernšteĭn [**10**], who obtained

$$N_k(\xi) \sim \frac{n^k}{[\xi(1-\xi)]^{k/2}} \text{ for } \xi \in [0,1];$$

it follows from this that

$$\lim_{n\to\infty} \frac{B_k(\xi)}{N_k(\xi)} = k^{k/2} \text{ for } \xi \in [0,1].$$

The best approximation to $N_k(\xi)$ is given by (18), which is tangent to $N_k(\xi)$ in each Čebyšev interval at $\xi = \vartheta_i^{(k)}$ ($k = 1, 2, \cdots, n$; $i = 1, 2, \cdots, n-k+1$), where $(\vartheta_i^{(k)})_{i=1}^{n-k+1}$ are the roots of $S_n^{(k)}(x) = 0$.

For example, if $k + 1 \equiv n \pmod 2$, we have for fixed k and large n

$$| T_n^{(k)}(\tfrac{1}{2}) + iS_n^{(k)}(\tfrac{1}{2}) | - N_k(\tfrac{1}{2}) \sim k2^k n^{k-1}$$

and

$$\frac{| T_n^{(k)}(\frac{1}{2}) + iS_n^{(k)}(\frac{1}{2}) |}{N_k(\frac{1}{2})} \sim 1.$$

I take this occasion to thank E. V. Voronovskaja for suggesting this problem and for her advice.

Bibliography

1. V. A. Markov, *On functions deviating least from zero in a given interval,* Izdat. Akad. Nauk, St. Petersburg, 1892. (Russian)

2. S. N. Bernšteĭn, *On V. A. Markov's theorem,* Trudy Leningrad. Indust. Inst. Razdel Fiz.-Mat. Nauk **5** (1938), no. 1, 8-13; reprinted in Collected works, Vol. 2, Izdat. Akad. Nauk SSSR, Moscow, 1954, pp. 281-286. (Russian) MR **16**, 433.

3. A. C. Schaeffer and R. J. Duffin, *On some inequalities of S. Bernštein and W. Markoff for derivatives of polynomials,* Bull. Amer. Math. Soc. **44** (1938), 289-297.

4. E. V. Voronovskaja, *Extremal polynomials of finite functionals,* Thesis Abstract, Leningrad. Gos. Univ., Leningrad, 1955. (Russian)

5. ———, *Application of functional analysis to polynomials of least deviation,* Dokl. Akad. Nauk SSSR **99** (1954), 5-8. (Russian) MR **17**, 842.

6. ———, *Extremal polynomials of some of the simplest functionals,* Dokl. Akad. Nauk SSSR **99** (1954), 193-196. (Russian) MR **17**, 842.

7. ———, *The functional of the first derivative and improvement of a theorem of A. A. Markov,* Izv. Akad. Nauk SSSR Ser. Mat. **23** (1959), 951-962. (Russian) MR **22** #1640.

8. E. I. Zolotarev, *Application of elliptic functions to problems on functions deviating least or most from zero,* Zapiski S.-Peterburg. Akad. Nauk **30** (1877), no. 5; reprinted in Collected works, Vol. 2, Izdat. Akad. Nauk SSSR, Leningrad, 1932, pp. 1-59. (Russian)

9. S. N. Bernšteĭn, *On the best approximation of continuous functions by polynomials of given degree,* Soobšč. Har'kov. Mat. Obšč. (2) **13** (1912), 49-191; reprinted in Collected works, Vol. I, Izdat. Akad. Nauk SSSR, Moscow, 1952, pp. 11-104. (Russian) MR **14**, 2.

10. ———, *Remarques sur l'inégalité de Wladimir Markoff,* Soobšč. Har'kov. Mat. Obsc. **14** (1913), 81-87; Russian transl., Collected works, Vol. I. Izdat. Akad. Nauk SSSR, Moscow, 1952, pp. 151-156. MR **14**, 2.

Index of Numbered Examples

Index of Numbered Theorems

SUBJECT INDEX